IONISATION CONSTANTS
OF INORGANIC ACIDS AND BASES
IN AQUEOUS SOLUTION

IUPAC CHEMICAL DATA SERIES

Number 1	Y. MARCUS	Critical Evaluation of Some Equilibrium Constants Involving Organophosphorus Extractants
Number 2	A.S. KERTES	Critical Evaluation of Some Equilibrium Constants Involving Alkylammonium Extractants
Number 3	Y. MARCUS	Eqilibrium Constants of Liquid-Liquid Distribution Reactions: Organophosphorus Extractants
Number 4	Y. MARCUS	Equilibrium Constants of Liquid-Liquid Distribution Reactions: Alkylammonium Salt Extractants
Number 5	S. ANGUS et al	International Thermodynamic Tables of the Fluid State: Argon
Number 6	S. ANGUS et al	International Thermodynamic Tables of the Fluid State: Ethylene
Number 7	S. ANGUS et al	International Thermodynamic Tables of the Fluid State: Carbon Dioxide
Number 8	S. ANGUS et al	International Thermodynamic Tables of the Fluid State: Helium
Number 9	A.R.H. COLE	Tables of Wavenumbers for the Calibration of Infrared Spectrometers, 2nd Edition
Number 10	Y. MARCUS and D.G. HOWERY	Ion Exchange Equilibrium Constants
Number 11	R. TAMAMUSHI	Kinetic Parameters of Electrode Reactions of Metallic Compounds
Number 12	D.D. PERRIN	Dissociation Constants of Organic Bases in Aqueous Solution
Number 13	G. CHARLOT et al	Selected Constants: Oxidation-Reduction Potentials of Inorganic Substances in Aqueous Solutions
Number 14	C. ANDEREGG	Critical Survey of Stability Constants of EDTA Complexes
Number 15	Y. MARCUS et al	Equilibrium Constants of Liquid-Liquid Distribution Reactions: Compound Forming Extractants, Solvating Solvents and Inert Solvents
Number 16	S. ANGUS et al	International Thermodynamic Tables of the Fluid State: Methane
Number 17	W.A.E. McBRYDE	A Critical Review of Equilibrium Data for Proton- and Metal Complexes of 1,10-Phenanthroline, 2,2'-Bipyridyl and Related Compounds
Number 18	J. STARY and H. FREISER	Equilibrium Constants of Liquid-Liquid Distribution Reactions: Chelating Extractants
Number 19	D.D. PERRIN	Dissociation Constants of Inorganic Acids and Bases in Aqueous Solution
Number 20	S. ANGUS et al	International Thermodynamic Tables of the Fluid State: Nitrogen
Number 21	E. HÖGFELDT	Stability Constants of Metal-Ion Complexes, Part A: Inorganic Ligands
Number 22	D.D. PERRIN	Stability Constants of Metal-Ion Complexes, Part B: Organic Ligands
Number 23	E.P. SERJEANT and B. DEMPSEY	Ionisation Constants of Organic Acids in Aqueous Solution
Number 24	J. STARY et al	Critical Evaluation of Equilibrium Constants Involving 8-Hydroxyquinoline and its Metal Chelates
Number 25	S. ANGUS et al	International Thermodynamic Tables of the Fluid State: Propylene (Propene)
Number 26	Z. KOLARIK	Critical Evaluation of Equilibrium Constants Involving Acidic Organophosphorus Extractants
Number 27	A.M. BOND and G.T. HEFTER	Critical Survey of Stability Constants and Related Thermodynamic Data of Fluoride Complexes in Aqueous Solution
Number 28	P. FRANZOSINI and M. SANESI	Thermodynamic and Transport Properties of Organic Salts
Number 29	D.D. PERRIN	Ionisation Constants of Inorganic Acids and Bases in Aqueous Solution, 2nd Edition

INTERNATIONAL UNION OF PURE AND APPLIED CHEMISTRY
(ANALYTICAL CHEMISTRY DIVISION, COMMISSION ON EQUILIBRIUM DATA)

IONISATION CONSTANTS OF INORGANIC ACIDS AND BASES IN AQUEOUS SOLUTION

Compiled by

D. D. PERRIN

Medical Chemistry Group
Institute of Advanced Studies
Australian National University, Canberra

Second Edition

IUPAC Chemical Data Series, No. 29

PERGAMON PRESS

OXFORD · NEW YORK · TORONTO · SYDNEY · PARIS · FRANKFURT

U.K.	Pergamon Press Ltd., Headington Hill Hall, Oxford OX3 0BW, England
U.S.A.	Pergamon Press Inc., Maxwell House, Fairview Park, Elmsford, New York 10523, U.S.A.
CANADA	Pergamon Press Canada Ltd., Suite 104, 150 Consumers Rd., Willowdale, Ontario M2J 1P9, Canada
AUSTRALIA	Pergamon Press (Aust.) Pty. Ltd., P.O. Box 544, Potts Point, N.S.W. 2011, Australia
FRANCE	Pergamon Press SARL, 24 rue des Ecoles, 75240 Paris, Cedex 05, France
FEDERAL REPUBLIC OF GERMANY	Pergamon Press GmbH, 6242 Kronberg-Taunus, Hammerweg 6, Federal Republic of Germany

First edition 1969
Second edition 1982

Library of Congress Cataloging in Publication Data

Perrin, D. D. (Douglas Dalzell), 1922-
Ionisation constants of inorganic acids and bases
in aqueous solution.
(IUPAC chemical data series ; no. 29)
Chiefly tables.
At head of title: International Union of Pure and
Applied Chemistry, Analytical Chemistry Division,
Commission on Equilibrium Data.
Rev. ed. of: Dissociation constants of inorganic
acids and bases in aqueous solution. 1969.
Bibliography: p.
1. Dissociation—Tables. 2. Acids, Inorganic—Tables.
3. Bases (Chemistry)—Tables. I. International Union
of Pure and Applied Chemistry. Commission on Equilibrium
Data. II. Title. III. Series.
QD561.P45 1982 541.3'722'0212 82-16524
ISBN 0-08-029214-3

In order to make this volume available as economically and as rapidly as possible the authors' typescripts have been reproduced in their original forms. This method unfortunately has its typographical limitations but it is hoped that they in no way distract the reader.

Printed in Great Britain by A. Wheaton & Co. Ltd., Exeter

CONTENTS

COMMISSION ON EQUILIBRIUM DATA

1979-1981

Titular Members

G. H. Nancollas (Chairman)
S. Ahrland (Secretary)
G. Anderegg, W. A. E. McBryde, H. Ohtaki,
D. D. Perrin, L. D. Pettit

Associate Members

D. S. Gamble, E. D. Goldberg, E. Högfeldt,
A. S. Kertes, P. W. Schindler, J. Stary, P. Valenta

National Representatives

A. F. M. Barton (Australia), M. T. Beck (Hungary),
A. Bylicki (Poland), I. N. Marov (USSR),
H. M. N. H. Irving (UK), A. E. Martell (USA)

1981-1983

Titular Members

S. Ahrland (Chairman)
H. Ohtaki (Secretary)
E. D. Goldberg, J. Grenthe, L. D. Pettit, P. Valenta

Associate Members

G. Anderegg, A. C. M. Bourg, D. S. Gamble,
E. Högfeldt, A. S. Kertes, W. A. E. McBryde, I. Nagypal,
G. H. Nancollas, D. D. Perrin, J. Stary, O. Yamauchi

National Representatives

A. F. M. Barton (Australia), M. B. Beck (Hungary),
A. Bylicki (Poland), C. Luca (Romania),
I. N. Marov (USSR), A. E. Martell (USA)

INTERNATIONAL UNION OF PURE AND APPLIED CHEMISTRY

IUPAC Secretariat: Bank Court Chambers, 2-3 Pound Way,
Cowley Centre, Oxford OX4 3YF, UK

PREFACE

These Tables have been compiled as part of the continuing work of the Commission on Equilibrium Data, Analytical Division, International Union of Pure and Applied Chemistry. They were published originally in *Pure and Applied Chemistry* (Volume 20, No. 2, 1969), and as a separate volume, *Dissociation Constants of Inorganic Acids and Bases in Aqueous Solution* (1969), by Butterworth and Co. Ltd., London.

As the Tables have been out of print for some years the opportunity has been taken in reprinting them, to update them to the end of 1980.

Most of the existing tables of ionisation constants of inorganic acids and bases in aqueous solution are fragmentary in character, include little or no experimental details, and give few references. Easily the most comprehensive of the previous collections is *Stability Constants of Metal-Ion Complexes*, compiled by L.G. Sillén and A.E. Martell, and published as Special Publication No. 17 of the Chemical Society, London, in 1964. However, because of the nature of this compilation, the p*K* values in it tend to be overlain by the much greater bulk of the stability constant data. In many cases, also, it is difficult to decide by inspection which of the p*K* values should be taken from the wide range sometimes given for a particular substance.

The present Tables follow the pattern of the similar Tables for organic acids and organic bases, which were also prepared at the request of the International Union of Pure and Applied Chemistry as part of the work of the Commission on Electrochemical Data. The Tables of organic acids, compiled by Kortum, Vogel, and Andrussow were published in *Pure and Applied Chemistry*, 1, 187-536 (1960), and also separately as a book[*]. These were revised and greatly expanded by E.P. Serjeant and Boyd Dempsey[ϕ]. The Tables of organic bases, by the present author, were published in 1965 as a supplement to *Pure and Applied Chemistry*, with a further volume in 1972[†].

For convenience, the ionisation constants of inorganic acids and bases have been given, in most cases, in the form of pK_a values, and the classes of compounds include not only conventional acids and bases such as boric acid and magnesium hydroxide, but also hydrated metal ions (which behave as acids when they undergo hydrolysis) and free radicals, such as the hydroxyl radical, .OH. All of these reactions have in common the gain or loss of a proton or a hydroxyl ion.

In general, and largely because of the difficulties attending p*K* measurements on inorganic species, it is not possible to offer a critical

* G. Kortum, W. Vogel and K. Andrussow. *Dissociation Constants of Organic Acids in Aqueous Solution*. Butterworth & Co. Ltd., London, 1961.

ϕ E.P. Serjeant and Boyd Dempsey, *Ionisation Constants of Organic Acids in Aqueous Solution* (IUPAC Chemical Data Series No. 23, Pergamon Press, Oxford, 1979.

† D.D. Perrin, *Dissociation Constants of Organic Bases in Aqueous Solution*, Butterworth & Co. Ltd., London, 1965; *Supplement*, 1972.

assessment of most of the published values. In particular cases, such as water and orthophosphoric acid, highly precise constants are available over a range of temperatures, and the uncertainty is only of the order of 0.001 pH unit. More commonly, only a few, often widely discordant, values have been reported.

This is partly because of the chemical reactivity of the materials themselves. For example, nitrous acid readily decomposes to dinitrogen trioxide. At concentrations above 0.01 M, boric acid is appreciably polymerised to polyboric acids; molybdic acid solutions contain $Mo_7O_{24}^{6-}$ and higher species; bisulphite ion is in equilibrium with pyrosulphite ion, $S_2O_5^{2-}$; and many transition and higher-valent metal ions form polynuclear species on hydrolysis.

Often, too, unsatisfactory methods of determination have been used. Thus, pH titration measurements are seldom satisfactory if p*K* values lie below 2 or above 12, and in such circumstances can give quite misleading results. Again, p*K* values for the hydrolysis of metal ions have often been obtained from measurements of the pH values of solutions of their purified salts in water. As Sillén has pointed out (*Quart. Rev.*, 13, 146 (1959), inorganic salts often adsorb tenaciously onto their surfaces traces of acidic or basic impurities, which persist even on repeated recrystallization, so that the measured pH values of their solutions may be much higher or lower than expected.

Even with experimentally accurate results, extrapolation to thermodynamic p*K* values at $I = 0$ is not always possible. The usual basis of such extrapolation is the Debye-Hückel equation,

$$-\log f\pm = \frac{z_i^2 A I^{\frac{1}{2}}}{1 + kaI^{\frac{1}{2}}} - bI$$

which is used to calculate the activity coefficient term. For precise work, values of a (the "mean distance of nearest approach" of the ions) and b are chosen to fit the data over a range of ionic strengths, so that the value of the p*K*, extrapolated to $I = 0$, can be obtained. At low ionic strengths and where moderate accuracy (say ±0.05 pH unit) is sufficient some simplifying assumptions can often be made. Thus, Davies' equation (*J. Chem. Soc.* 1938, 2093) is obtained by taking $Ka = 1$, $b = 0.2$; Güntelberg's equation (*Z. physik. Chem. Leipzig*, 123, 199 (1926) sets $Ka = 1$, $b = 0$; and the approximation $Ka = 0$, $b = 0$ (*i.e.* $-\log f = z_i^2 AI^{\frac{1}{2}}$) is also used. However, with moderately strong acids and bases (p*K* values less than 2 or greater than 12), the numerical values of the thermodynamic p*K* constants depend in part on the assumptions made in deriving them, including the ion-size parameter a used in the extended Debye-Hückel equation (see, example, R.G. Bates, V.E. Bower, R.G. Canham and J.E. Prue, *Trans. Faraday Soc.*, 55. 2062 (1959); A.K. Covington, J.V. Dobson and W.F.K. Wynne-Jones, *Trans. Faraday Soc.*, 61, 2057 (1965), E.A. Guggenheim, *Trans. Faraday Soc.*, 62, 2750 (1966). Thus, the p*K* of bisulphite ion at 25° varies from 1.927 to 1.967 as K_a is varied from 1.0 to 1.7. In the same way, pK_b for $Ca(OH)_2$ varies from 1.14 to 1.27 at 25°, depending on the choice of parameters.

A distinction must also be made between true and apparent p*K* values. The first p*K* of carbon dioxide in water as measured is about 6.4 at 20°, whereas the true p*K* of carbonic acid (H_2CO_3) is 3.8. The difference between the apparent and the true p*K* values is due to the slight extent to which carbon dioxide is covalently hydrated in water. Similarly, periodic acid exists as H_5IO_6 and HIO_4 (mainly as the latter), so that its measured second p*K* (8.3) is very much higher than its first one (about 2).

In the absence of experimental values, especially for some of the oxyacids, attempts have been made to predict p*K* values, usually from similarities of structure. The more commonly used methods are those of J.E. Ricci (*J. Am. Chem. Soc.*, 70, 109 (1948), L. Pauling (*General Chemistry*, Freeman, San Francisco, 1947, p. 394), and A. Kossiakoff and D. Harker (*J. Am. Chem. Soc.*, 60, 2047, (1938)). Even in apparently simple cases, there may be considerable uncertainty. For example different values would be predicted for germanic acid depending on whether it existed mainly as $GeO(OH)_2$ or $Ge(OH)_4$.

Because of the many different kinds of uncertainties inherent in the present p*K* compilation, no attempt has been made to assess the accuracy of each entry. Nevertheless, where possible, I have attempted to select what appear to be the best available values. The results for hydrogen sulphide illustrate this. Thus, several methods have indicated that the second p*K* of hydrogen sulphide is about 14, which is too high for potentiometric titration methods to be applicable. Hence the pK_2 values that have been obtained by potentiometric titration are not set out in this Table. Instead, references to the papers where they are given are included under "other measurements". This heading also covers results where insufficient experimental details are given.

HOW TO USE THE TABLES
GENERAL ARRANGEMENT

The Tables summarize data recorded in the literature up to the end of 1980 for the ionisation constants of inorganic acids and bases in aqueous solution. They also include references to acidity functions for strong acids and bases, and details about the formation of polynuclear species where this is relevant. The substances are listed alphabetically, with chemical formulae, so that the entries are self-indexing.

Column 1 gives the name of the substance and the negative logarithm of the ionisation constant (pK_a). Wherever possible, these values are thermodynamic ones obtained by extrapolation to ionic strength $I = 0$, generally by using some form of the Debye-Hückel equation such as that due to Davies. In all cases, pK values are listed in decreasing extent of protonation.

Column 2 gives the temperature of measurements in ^{o}C.

Column 3 lists details such as:

$I = \frac{1}{2} \Sigma C_i Z_i^2$ = ionic strength

c = concentration in mole/l, or

m = concentration in mole/1000 g. of water.

It also records any other details relating to the pK value quoted. Designation of a constant as "practical" implies that it includes both the activity of the hydrogen ion (usually as measured by pH meter) and the concentrations of the other species.

Column 4 summarises the method of measurement, the procedure used in evaluating the constants, and any corrections that were taken into consideration; the symbols have the meanings set out under "Methods of Measurement". Because different investigators rarely use identical procedures, these symbols can only serve as guides: for fullest details the original papers should be consulted.

Column 5 gives the literature references which are listed alphabetically at the end of the Tables.

METHODS OF MEASUREMENT AND CALCULATION

The abbreviations in *Column 4* of the Tables are, with only minor differences, the same as those used in "Dissociation Constants of Organic Bases in Aqueous Solution".

CONDUCTOMETRIC METHODS

C1 Measurements in solutions of salt and acid

C2 Measurements in solution of base only

ELECTROMETRIC METHODS

[i] Cells without diffusion potentials

E1a Method of Harned and Ehlers (*J. Am. Chem. Soc.* 54, 1350 (1932)) (Cell of type Pt (H_2)B, BCl, NaCl‖B, BCl, NaCl|AgCl|Ag, for which $E - E_0 + (RT/F)$ ln $[BH^+]$ $[Cl^-]/[B] = -(RT/F)$ ln K', and extrapolate to $I = 0$)

E1b Method of Harned and Owen (*J. Am. Chem. Soc.* 52, 5079 (1930)), Pt(H_2)B, NaCl|AgCl|Ag, where molality of B is M, $E = E_0 - (RT/F)$ ln $([m_{H}{}^{+}]$ $[m_{Cl}{}^{-}]f{\pm}^{2}0$. Extrapolate to $I = 0$ at constant M, then to $M = 0$)

E1cg Determination of $[H^+]$ from cells of the type, Glass/solution, Cl^-|AgCl|Ag

E1ch Determination of $[H^+]$ from the cell, Pt(H_2) solution, Cl^-| AgCl|Ag

E1d Method of Bates (*J. Am. Chem. Soc.* 70, 1579 (1948)). Determination of K_1 and K_2 for dibasic acids

E1e Method of Bates and Pinching (*J. Res. Natl. Bur. Std.* 43, 519 (1949)). A particular case of method E1cg in which the solution is a buffer comprising a weak base and a weak acid

[ii] Approximately symmetrical cells with diffusion potentials

E2a Method of Owen (*J. Am. Chem. Soc.* 60, 2229 (1938))

E2b Method of Larsson and Adell (*Z. Physik. Chem.* 156, 352, 381 (1931)) (Uses cell Pt(H_2)|B, NaCl|sat. KCl|NaOH, NaCl|(H_2)Pt and an approx. K to adjust to equal ionic strengths in the half-cells. From E obtain $[H^+]$ and hence K': extrapolation to I = 0 gives K)

E2c Method of Everett and Landsman (*Proc. Roy. Soc. London*, A215, 403, (1952))
(This is like E2b but uses a second weak base of known pK instead of a strong base. The method gives the ratio of the two constants)

[iii] Unsymmetrical cells with diffusion potentials

E3ag pH measurements in buffer solutions of weak electrolytes using glass electrodes

E3ah Similar measurements using hydrogen electrodes

E3bg Measurements of pH changes during titrations using glass electrodes

E3bh Similar measurements using hydrogen electrodes

E3b, quin Similar measurements using quinhydrone electrodes

E3c Differential potentiometric methods

E3d pH measurements at equal concentrations of salt and base

OPTICAL METHODS

01 Direct determination of the degree of dissociation by extinction coefficient measurements in solutions of weak bases and salts

02 Colorimetric determination with an indicator of known pK

03 Colorimetric determination with an indicator calibrated with a buffer solution of known pH

04 Method of von Halban and Brüll (*Helv. Chim. Acta* 27, 1719 (1944)) (Solutions of the base being studied, plus indicator, are compared with similar solutions containing alkali and indicator)

05 Light absorption measurements combined with electrometric measurements

06 Light absorption measurements using solutions of mineral acids of known concentrations and (usually) Hammett's acidity function, H_0

07 Similar to 06 using solutions of alkalis

OTHER METHODS

ANALYT Constants derived from chemical analysis

CALORIM Calorimetric measurements

CAT Constants estimated from catalytic coefficients

CRYOSC Cryoscopic measurements

DISTRIB	Distribution between solvents
FP	Constants derived from freezing-point data
ION	Ion-exchange studies
KIN	Constants estimated from kinetic measurements
NMR	Nuclear magnetic resonance measurements
POLAROG	Polarographic measurements
RAMAN	Measurements of Raman spectra
REDOX	Oxidation-reduction potentials
SOLY	Solubility measurements
VAP	Vapour pressure measurements

CALCULATIONS

[i] Conductance measurements

R1a Method of Davies (*The Conductivity of Solutions*, Chapman Hall, London 1930)
(By successive approximations, f_Λ is calculated from the Debye-Hückel-Onsager equation in the form

$$f_\Lambda = 1 - A(\alpha c_0)^{\frac{1}{2}}/\Lambda_0$$

which assumes that Λ_0 can be obtained from Kohlrausch's law of independent ionic mobilities)

R1b Method of MacInnes (*J. Am. Chem. Soc.* 48, 2068 (1926))
(The quantity $\Lambda_e = f_\Lambda \Lambda_0$ is determined directly, where Λ_e is the conductance of the weak electrolyte if it were completely dissociated at the ionic strength studied: it is necessary to know Λ for strong electrolytes as a function of I)

R1c Method of Fuoss and Krauss (*J. Am. Chem. Soc.* 55, 476 (1933))
(The Debye-Hückel-Onsager equation is used in the form, $\Lambda_c = \alpha(\Lambda_0 - \Lambda(\alpha c_0)^{\frac{1}{2}})$ to derive an equation relating Λ_0, c and K, which is solved by successive approximation until Λ_0 is constant at all values)

R1d Method of Shedlovsky (*J. Franklin Inst.* 225, 739 (1938))
(This is like R1c but a different equation is used)

R1e Method of Fuoss (*J. Am. Chem. Soc.* 79, 3301 (1957))

[ii] Differential potentiometric measurements

R2a Method of Kilpi (*Z. Physik. Chem.* 173, 223, 427 (1935); 175, 239 (1936) (at point of inflection).

Name, Formula and pK value	T(^{o}C)	Remarks	Methods	Reference
1. (Aquo) **Aluminium ion**, Al^{3+}				
4.89, 5.43, 5.86	25	I = 0.1 Successive pK values for hydrolysis of Al^{3+}	08	N24
4.9	25	I = 0	C	N42
5.04, 5.61, 6.10	25	I = 0.1 – 1.0 ($NaClO_4$), c = 0.01M; extrapolated to I = 0	08	N20
4.31	25	I = 1 ($NaClO_4$)	KIN	S105
5.28	15	pK for hydrolysis of Al^{3+}, c = 0.0005 – 0.01 M in $AlCl_3$, extrapolated against $I^{1/2}$	E3ag	S32
5.15	20			
4.98	25			
4.96	25	pK for hydrolysis of Al^{3+}, I varied from 0.0025 to 0.019, extrapolated to I = 0	E3ag	H46
5.02	25	pK for hydrolysis of Al^{3+}, c = 10^{-5} – 10^{-2} M in $Al(ClO_4)_3$, extrapolated to I = 0	E3ag	F48
4.96	25	pK for hydrolysis of Al^{3+}; also log K = 7.55 for $2AlOH^{2+} \rightleftharpoons Al_2(OH)_2{}^{4+}$, and log K = 6.89 for $2Al_2(OH)_4{}^{2+} + 2H_2O \rightleftharpoons Al_4(OH)_{10}{}^{2+} + 2H^+$	E3,quin	K23
5.03	25	pK for hydrolysis of Al^{3+}	E3ag	K100
5.10	25	pK for hydrolysis of Al^{3+}	E3ag	I21
4.49	25	pK for hydrolysis of Al^{3+}, from dissociation field effect relaxation times	KIN	H80
2.88	100	pK for hydrolysis of Al^{3+}	KIN	K108
11.22	25	pK for $Al(OH)_3 + H_2O \rightleftharpoons Al(OH)_4{}^- + H^+$	C1	M18
		Hydrolysis of Al^{3+} in 2 M $NaClO_4$ at 40° gives, mainly, one or more polynuclear complexes		B131
7.07	25	I = 2(Na)ClO_4; $-\log {}^*\beta_{22}$		A24
7.07	25	I = 1($NaClO_4$); $-\log {}^*\beta_{22}$; ultracentrifugation	OTHER	A59
6.95	25	I = 0; $-\log {}^*\beta_{22}$	KIN	T32

Name, Formula and pK value	T(^{o}C)	Remarks	Methods	Reference
7.45	30	$\underline{I}$ = 0; $-\log^{*}\beta_{22}$; pH titration		G62
		For polynuclear complex formation by hydrolysed Al^{3+}, see M61		
		For equilibria between $Al(OH)_3$ and OH^-, see P9		
		Other Measurements: B128, D36, F6, I25, K67, L1, T2, T10, V19, W33		
2. (Aquo) Americium (III) ion Am^{3+}				
5.92	23	I = 0.1 ($HClO_4$, $LiClO_4$); p$\underline{K}$ for hydrolysis of Am^{3+}	DISTRIB	D46
10.7	15	I = 0.005 (Cl, KCl); paper electrophoresis	OTHER	M31
		Other measurements: K74		
3. Amidophosphoric acid $NH_2PO_3H_2$				
2.610 8.422	6.0	I = 0.5, extrapolated to $\underline{I}$ = 0	E3bg	P23
2.708 8.320	9.2			
2.738 8.184	19.0			
2.739 8.102	26.5			
2.731 7.950	33.0			
2.716 7.896	40.0			
2.696 7.821	46.3			
3.00 8.15	25	$\underline{I}$ = 0.2 (KCl), "practical" constants	E3bg	C17a
3.3 8.28	25	$\underline{I}$ = 1(NMe_4Br), concentration constants, $f_{\pm}$ assumed same as for HBr	E3bg	I13
3.8 10.25		Titration of 0.1 M solution; p$\underline{K}$ of $^{+}NH_3PO_3H_2$ given as 2.1	E3b	R54
2.92 7.86	20		E,Sb	K40
2.8 8.2			E,h	M69
4.6 7.7	10			H15
8.2	25			R13

			Other measurements: C23		
3.08	8.63	25	$I \rightarrow 0$	E3bg	L27

4. **Amidotriphosphoric acid**, $NH_2P_3O_9H_4$

5.8		20	pK_4	E3bg	F17

5. **Aminodisulphonic acid**, $NH(HSO_3)_2$

8.50		25	pK_3; $I = 1.0$ (NaCl)	E3ag	D58

6. **Aminophosphazenes**, see Hexaminotriphosphazene, Octaminotetraphosphazene.

7. **Aminophosphoric acid**, see Amidophosphoric acid.

8. **Ammonia**, NH_3

10.081	0	Equal concentrations of NH_3 and KH phenol sulphonate, varied from 0.011 to 0.104 M, activity coefficients calculated from Debye-Huckel equation, pK plotted against I	E1ch	B24
9.903	5			
9.730	10			
9.564	15			
9.401	20			
9.246	25			
9.093	30			
8.947	35			
8.805	40			
8.671	45			
8.540	50			
10.081	0		E1a	B23

Name, Formula and p*K* value	T(°C)	Remarks	Methods	Reference
9.904	5			
9.731	10			
9.564	15			
9.400	20			
9.245	25			
9.093	30			
8.947	35			
8.805	40			
8.670	45			
8.539	50	Thermodynamic quantities are derived from these values.		
9.555	15	*I* varies from 0.06 to 0.20. Extrapolated to zero	E2b	E35
9.240	25	concentration of NH_{4+} at each *I*, then extrapolated		
8.946	35	against *I*.		
8.670	45	Thermodynamic quantities are derived from these values.		
10.070	0	*I* = 0	CAL	O17
9.241	25			
8.536	50			
7.928	75			
7.400	100			
6.935	125			
6.523	150			
5.824	200			
5.253	250			
4.776	300			
9.867	5	*c* = 0.02 to 0.08; *I* = 0.07 to 0.2, extrapolated to	E2b	E37
9.529	15	*I* = 0, *c* = 0		
9.215	25			

8.923	35			
8.645	45			
10.19	0		Cl	N46
9.58	18			
9.35	25			
7.45	100			
6.45	156			
5.60	218			
5.74	306			
4.68	49	pK_b values	Cl	W39
4.83	93			
5.04	138			
5.36	182			
5.76	227			
6.21	271			
6.62	293			
4.862	0	pK_b; recalculation of Literature values	LIT	F25
4.830	5			
4.804	10			
4.782	15			
4.766	20			
4.752	25			
4.740	30			
4.734	35			
4.730	40			
4.726	45			
4.723	50			
4.76	75			
4.84	100			
4.96	125			

Name, Formula and p$\underline{K}$ value	T(^{o}C)	Remarks	Methods	Reference
5.11	150			
5.47	200			
5.68	225			
5.91	250			
6.17	275			
6.47	300			
6.83	325			
7.30	350			
4.864	0	$\underline{I}$ = 0.01–0.20 (KCl); extrapolated to $\underline{I}$ = 0; p$\underline{K}_b$	E3bh	H14
4.752	25	Thermodynamic quantities are derived from the results		
4.732	50			
4.772	75			
4.856	100			
4.976	125			
5.128	150			
5.311	175			
5.525	200			
5.770	225			
6.047	250			
6.355	275			
6.694	300			
7.58	100	taking p$\underline{K}_w$ = 12.38; inversion of sucrose	CAT	K108
		Ref. H56 gives an equation fitting literature values of p$\underline{K}$ from 0° to 300°.		
4.29	25	p$\underline{K}_b$ values 1000 atmospheres pressure	Cl,Rla	B139
3.91	25	2000		
3.61	25	3000		
4.32	45	1000		
3.95	45	2000		

3.65	45	3000		
4.71	45	pK_b values 1 atmosphere pressure	Cl,Rla	H17
4.30		1100		
3.74		2500		
3.32		4000		
2.95		5400		
2.68		6800		
2.42		8200		
2.21		9600		
2.11		11000		
2.00		12000		
32.72	-50	Self-ionization of liquid ammonia, from cell potential data		P49
32.49	-33.2	Self-ionization of liquid ammonia, from thermodynamic data		C48
27.66	24.8			
29.8	25	Self-ionization of liquid ammonia, from thermodynamic data		J15
40		Approximate pK of NH_2^-, theoretical calculation		S48
		A value of 4.20 at 25° has been claimed from high field conductance measurements to be the true pK_b of $NH_4^+ + OH^- \rightleftharpoons NH_4OH$		B47
		A similar value, 4.28 at 20°, has been estimated from published data		M93
		For pK values in methanol-water mixtures, see A13, E38, P1		
		Other measurements: A13, B76, F57, H22, H31, H42, K9, K37, L64, M94, N47, O24, P25, Q6, S50, W30		
16.46	25	pK in CH_3CN	E,g	C38
10.45	25	$I = 10(NH_4NO_3)$	E,g	L51
		pK values of some metal-ammine complexes		
5.15	20	pK for $[Cr(NH_3)_5(H_2O)]^{3+}$; $I = 0.1$	E3bg	C
5.58		pK in D_2O		

Name, Formula and p$\underline{K}$ value	T(^{o}C)	Remarks	Methods	Reference
5.25		p$\underline{K}$ in 20% dioxane/water		
6.18	20	p$\underline{K}$ for $[Co(NH_3)_5(H_2O)]^{3+}$; $\underline{I}$ = 0.1	E3bg	C
6.67		p$\underline{K}$ in D_2O		
6.30		p$\underline{K}$ in 20% dioxane/water		
5.69		p$\underline{K}$ for $[Co(NH_3)_5(H_2O)]^{3+}$	LIT	B
5.22		p$\underline{K}$ for $[Co(NH_3)_4(H_2O)_2]^{3+}$	LIT	B
4.73		p$\underline{K}$ for $[Co(NH_3)_3(H_2O)_3]^{3+}$	LIT	B
3.40		p$\underline{K}$ for $[Co(NH_3)_2(H_2O)_4]^{3+}$	LIT	B
9.5, 10.7		stepwise p$\underline{K}$ value for $[Pt(NH_3)_5OH]^{3+}$	LIT	B
8.2, 10.4		stepwise p$\underline{K}$ values for $[Pt(NH_3)_5Br]^{3+}$	LIT	B
8.1, 10.5		stepwise p$\underline{K}$ values for $[Pt(NH_3)_5Cl]^{3+}$	LIT	B
7.9, 10.1		stepwise p$\underline{K}$ values for $[Pt(NH_3)_6]^{4+}$	LIT	B
10.9		p$\underline{K}$ for $[Pt(NH_3)_4NH_2Cl]^{2+}$	LIT	B
9.8		p$\underline{K}$ for $(Pt(NH_3)_4Cl_2]$	LIT	B
5.56, 7.32		stepwise p$\underline{K}$ value for <u>cis</u> $[Pt(NH_3)_2(H_2O)_2]^{2+}$		
5.63, 9.25		stepwise p$\underline{K}$ value for <u>cis</u> $[Pt(NH_3)_2(H_2O)_2]^{2+}$		
5.86		p$\underline{K}$ for $[Rh(NH_3)_5H_2O]^{3+}$	LIT	B
6.93	25	p$\underline{K}$ for $[Rh(NH_3)_5H_2O]$ $\underline{I}$ = 1($NaClO_4$)		S83a
6.40, 8.32	25	stepwise p$\underline{K}$ for $[Rh(NH_3)_4(H_2O)Cl^-]$; cis	E	S83a
4.92, 8.26	25	stepwise p$\underline{K}$ for $[Rh(NH_3)_4]$; trans	E	S83a
7.84	25	p$\underline{K}$ for $Rh(NH_3)_4(Cl)(OH)$; cis	E	S83a
6.75	25	p$\underline{K}$ for $Rh(NH_3)_4(Cl)(OH)$; trans	E	S83a
7.89, 6.87	25	p$\underline{K}$ for $Rh(NH_3)_4(OH)Br$; cis	E	S83a

B. F. Basolo, Chap. 10, "Acids, Bases and Amphoteric Hydroxides" in "The Chemistry of the Coordination Compounds", Ed. J.C. Bailar and D.A. Busch, Reinhold Publ. Co., New York, 1956.

C. S.C. Chan and K.Y. Hui, <u>Austral. J. Chem.</u> **21** 3061 (1968)

9. (Aquo) Antimony (III) ion, Sb^{3+}

1.42	23	$\underline{I}$ = 1($HClO_4$); p$\underline{K}$ for $SbO^+ + H_2O \rightleftharpoons Sb(OH)_3 + H^+$		M78
11.0	25	p$\underline{K}$ for $Sb(OH)_3 + OH^- \rightleftharpoons Sb(OH)_4^- + H^+$	O8	A42
12.95		log K for $Sb(OH)_3 \rightleftharpoons Sb(OH)_2^+ + OH^-$		
1.4	25	p$\underline{K}$ for $SbOH^{2+} \rightleftharpoons SbO^+ + H^+$	SOLY	K12
11.8	25	p$\underline{K}$ for $SbO^+ + H_2O \rightleftharpoons HSbO_2 + H^+$	SOLY	
0.87	25	p$\underline{K}$ for $SbO^+ + H_2O \rightleftharpoons HSbO_2 + H^+$	SOLY	P48
11.0	25	p$\underline{K}$ for $HSbO_2 + 2H_2O \rightleftharpoons Sb(OH)_4^- + H^+$		

10. Antimony pentoxide, Sb_2O_5 See also Dodeca-antimonic acid.

2.55	25	p$\underline{K}$ for $HSb(OH)_6 \rightleftharpoons Sb(OH)_6^- + H^+$; $\underline{I}$ = 0.5(NMe_4Cl); Sb concentration $\leqslant 10^{-3}$M; at higher concentrations polynuclear complexes are also formed.	E3b	L21

Aquo metal ion, See entry under appropriate metal ion

11. Arsenic acid, H_3AsO_4

2.089	7.054		0	$\underline{I}$ varied from 0.007 to 0.096 (for $\underline{K}_1$).	Ela,quin	A14
2.114	7.032		5	and 0.010 to 0.21 (for $\underline{K}_2$);		
2.138	7.015		10	extrapolated to $\underline{I}$ = 0		
2.163	6.999		15			
2.194	6.990		20			
2.223	6.980		25			
2.265	6.974		30			
2.296	6.973		35			
2.332	6.973		40			
2.26	6.76	11.29	25	$\underline{I}$ = 0.1(KCl)	E3bg	T21
2.30			25	$\underline{I}$ = 0	E3bg	S6

Name, Formula and p$\underline{K}$ value	T(^{o}C)	Remarks	Methods	Reference
2.383 6.973	45			
2.420 6.980	50			
$p\underline{K}_1 = 2.014 + 5 \times 10^{-5}\ (t - 40.0)^2$				
$p\underline{K}_2 = 6.971 + 5 \times 10^{-2}\ (t - 39.4)^2$, t in ^{o}C.				
		Thermodynamic quantities are derived from the results.		
2.49 7.05 11.33	10		E,g	F31
2.19 6.94 11.50	25			
1.95 6.87 11.64	35			
2.15 6.80 11.92	50			
7.08	25	Taking p$\underline{K}_2$ of H_3PO_4 as 7.16	E3ag	H98
2.301	25		E3ag	S8
$p\underline{K}_2 = 306/\underline{T} + 5.925$, where $\underline{T}$ is in ^{o}K			LIT	E6
For p$\underline{K}_1$ in CH_3OH, C_2H_5OH or dioxane/water see T21				
		For values of p$\underline{K}_1$ in D_2O/H_2O mixtures, see S8		
		Other measurements: B83, B120, C32, K72a, L65, M14, S87, W7, W8		
12. Arsenious acid, $H_2AsO_3(HAsO_2)$				
9.295	15	Molal scale; c = 0.008, $\underline{I}$ = 0.1 (KCl)	E3dg	A40
9.265	20			
9.18	25			
9.09	30			
8.97	35			
8.885	40			
8.81	45			
9.294	25	In KCl solutions, extrapolated to $\underline{I}$ = 0	E	A39
9.22	25	Taking p$\underline{K}$ of boric acid as 9.19	E3ag	H98
9.26	18	"Practical" constant, titration of 0.017 M H_2AsO_3	E3bg	B120
9.08	25		E3ag	I16

9.4	13.5	Room	$p\underline{K}_2$ obtained from ultraviolet spectra	O	G51
	13.8	32		CRYOSC	S98
			Other measurements: B83, C13, G11, K52, K72a, T22, W8, W33, Z2		
9.28		25	Extrapolated to $\underline{I} = 0$	E3bg	S6
9.32		22	$p\underline{K}$ for $H_3AsO_2 \rightleftharpoons H^+ + HAsO_2^-$; $\underline{I} = 0.5$ (NaCl)	E,O	I23
9.21			$\underline{I} = 1.0$		
			Self association also occurs		
			For variation of $p\underline{K}_a$ in D_2O/H_2O mixture, see S6		
13.54,	13.99	20	$\underline{I} = 1$ (NaCl); $p\underline{K}_2$, $p\underline{K}_3$	E	S22
13.65,	13.75	20	$\underline{I} = 1$ (KCl); $p\underline{K}_2$; $p\underline{K}_3$	O	
44.9			$0.1^-3\underline{M}$ $HClO_4$; log β_3 for $3H^+ + AsO_3^{3-} \rightleftharpoons H_3AsO_3$	DISTRIB	L26

13. Azido-dithiocarbonic acid, $HSCSN_3$

1.67		25	Free acid readily decomposes	C1	S89

14. (Aquo) Barium ion, Ba^{2+}

0.62		5	$p\underline{K}_b$ of $BaOH^+$; $\underline{I} = 0.1$; f± calculated by Davies' equation,		
0.60		15	for extrapolation to $\underline{I} = 0$; from e.m.f. data of H.S. Harned		
0.64		25	and C.G. Geary, J. Am. Chem. Soc., 59 2032 (1937)		
0.69		25			
0.72		45			
			Thermodynamic quantities are derived from the results.		
0.64		25	$p\underline{K}_b$ of $BaOH^+$; $\underline{I} = 0.04$ to 0.17; using Davies' equation and activity measurements of H.S. Harned and C.M. Mason, J. Am. Chem. Soc. 54, 1441 (1932)		D17
0.85		25	$c = 0.02 - 0.05$ ($Ba(OH)_2$), $\underline{I} = 0.23$ to 0.6 ($Ba(OH)_2$ + $BaCl_2$), extrapolation to $\underline{I} = 0$, using Davies' equation	KIN	B39
0.62		25	$\underline{I} = 0.1$ to 0.45	CATKIN	B40
0.72			Concentration constant; 0.2 – 1 N $BaCl_2$; salt effect on	O3	K55

Name, Formula and p*K* value	T(°C)	Remarks	Methods	Reference
		indicator		
0.00	25	*I* =3 ($NaClO_4$) Other measurements: B41, K93	E2ah	C7
15. (Aquo) Berkelium (III) ion, Bk^{3+}				
5.66	23	p*K* for hydrolysis of Bk^{3+}; *I* = 0.1 ($HClO_4$ + $LiClO_4$)	DISTRIB	D46
16. (Aquo) Beryllium ion, Be^{2+}				
		Beryllium ions readily hydrolyze in solution and form condensed species containing more than one beryllium atom. See, for example, C8 and K6.		
5.7 ~7	20	Successive p*K* values for hydrolysis of Be^{2+}; *I* = 0.1 ($NaClO_4$); rapid-reaction measurements: $BeOH^+$ quickly forms trimer $Be_3(OH)_3^{3+}$	E3ag	S49
6.5	25	p*K* for $Be^{2+} \rightleftharpoons BeOH^+ + H^+$; *I* = 1($NaClO_4$); Be_2OH^{3+} also formed	E3bg	M49
>6.1	25	p*K* for $Be^{2+} \rightleftharpoons BeOH^+ + H^+$; *I* = 3($NaClO_4$); recalculation of data from refs. C8 and K6 using a computer; also −log *K* = 10.87 for $Be^{2+} + 2H_2O \rightleftharpoons Be(OH)_2 + 2H^+$; constants given for $Be_3(OH)_3^{3+}$ and Be_2OH^{3+}		H67
10.82		p*K*b for $Be(OH)_2 \rightleftharpoons BeOH^+ + OH^-$; *c* = 0.01; between pH 6.2 – 5.4; at lower pH values di- and tri-nuclear complexes are formed; constants are given	E3b	A11
10.46		p*K* for $Be(OH)_2 + H_2O \rightleftharpoons Be(OH)_3^- + H^+$; tracer concentrations also −log *K* = 13.65 for $Be^{2+} + 2H_2O \rightleftharpoons Be(OH)_2 + 2H^+$	DISTRIB	G59
3.16	25	*I* = 3($NaClO_4$); *c* = 2.5 −10 mM in Be^{2+}; −log *K* for $2Be^{2+} \rightleftharpoons Be_2OH^{3+} + H^+$		K1
8.66		−log *K* for $3Be^{2+} \rightleftharpoons Be_3(OH)_3^{3+} + 3H^+$		

11.16		$-\log \beta_2$		
3.28		in D_2O; $-\log \underline{K}$ for $2Be^{2+} \rightleftharpoons Be_2OD^{3+} + D^+$		
9.40		$-\log \underline{K}$ for $3Be^{2+} \rightleftharpoons Be_3(OD)_3^{3+} + 3D^+$		
11.89		$-\log \beta_2$		
		For variation of $\log \beta_{33}$ with D/H ratio, see K4		
3.27	25	$\underline{I} = 3(LiClO_4)$; coulometric titration; $-\log \underline{K}$ for Be_2OH^{3+}	OTHER	O8
11.5		$-\log \beta$ for $Be(OH)_2$		
8.74		$-\log \underline{K}$ for $Be_3(OH)_3^{3+}$		
3.66	25	$\underline{I} = 3(LiClO_4)$; 0.2 mole fraction dioxan-water, 2.5 - 80 mM Be^{2+}; $-\log \underline{K}$ for Be_2OH^{3+}	E	O5
10.8		$-\log \beta$ for $Be(OH)_2$		
7.15		$-\log \underline{K}$ for $Be_2(OH)_2$		
8.75		$-\log \underline{K}$ for $Be_3(OH)_3$		
10.08	0	$-\log \underline{K}$ for $3Be^{2+} \rightleftharpoons Be_3(OH)_3^{3+} + 3H^+$; $\underline{I} = 1(NaCl)$	E,H,quin	M61
8.91	25			
7.67	60			
3.64	0	$-\log \underline{K}$ for $2Be^{2+} \rightleftharpoons Be_2OH^{3+} + H^+$; $\underline{I} =$ 1(NaCl)	E,H,quin	
3.43	25			
2.93	60			
3.27	25	$-\log \underline{K}$ for $2Be^{2+} + H_2O \rightleftharpoons Be_2OH^{3+} + H^+$; $\underline{I} = 3(LiClO_4)$	E3bg	O8
8.74	25	$-\log \underline{K}$ for $3Be^{2+} + 3H_2O \rightleftharpoons Be_3(OH)_3^{3+} + H^+$; $\underline{I} = 3(LiClO_4)$		
11.5	25	$-\log \underline{K}$ for $Be^{2+} + 2H_2O \rightleftharpoons Be(OH)_2 + 2H^+$; $\underline{I} = 3(LiClO_4)$		
3.28	25	$-\log \underline{K}$ for $2Be^{2+} + H_2O \rightleftharpoons Be_2OH^{3+} + H^+$; $\underline{I} = 2(KNO_3)$	E3bg	L10
		Also $-\log \beta_{33} = 8.90$; $-\log \beta_{43} = 16$; $-\log \beta_{86} = 2.75$;		
		$-\log \beta_{96} = 34.5$		
3.18	25	$-\log \underline{K}$ for $2Be^{2+} + H_2O \rightleftharpoons Be_2OH^{3+} + H^+$; $\underline{I} = 3(KCl)$		P14
		Also $-\log \beta_{33} = 8.91$		
2.9	60	$-\log \underline{K}$ for $2Be^{2+} + H_2O \rightleftharpoons Be_2OH^{3+} + H^+$; $\underline{I} = 3(NaClO_4)$, $c = 1.25.10^{-3}$ to 0.05 $\underline{M}$		C34

Name, Formula and p$\underline{K}$ value	T(^{o}C)	Remarks	Methods	Reference
6.25	60	$-\log \underline{K}$ for $2Be^{2+} + 2H_2O \rightleftharpoons Be_2(OH)_2^{2+} + 2H^+$		
7.7	60	$-\log \underline{K}$ for $3Be^{2+} + 3H_2O \rightleftharpoons Be_3(OH)_3^{3+} + 3H^+$		
13.2	60	$-\log \underline{K}$ for $3Be^{2+} + 4H_2O \rightleftharpoons Be_3(OH)_4^{2+} + 4H^+$		
		$-\log \underline{K}$ = 8.81 for $3Be^{2+} \rightleftharpoons Be_3(OH)_3^{3+} + 3H^+$; $-\log \underline{K}$ = 3.24 for $2Be^{2+} \rightleftharpoons Be_2OH^{3+} + H^+$; $-\log \underline{K}$ = 11.0 for $Be^{2+} \rightleftharpoons Be(OH)_2 + 2H^+$; all for $\underline{I}$ = 0.5($NaClO_4$), *c* 0.001 to 0.08 $\underline{M}$ in Be^{2+}		B49
		$-\log \underline{K}$ = 10.9 for $Be^{2+} \rightleftharpoons Be(OH)_2 + 2H^+$, at 25° and $\underline{I}$ = 3($NaClO_4$); constants also given for di- and tri-nuclear species.		K6
		Other measurements: L53, T28, W34.		
17. (Aquo) <u>Bismuth (III) ion</u>, Bi^{2+}				
1.58	25	p$\underline{K}$ for $Bi^{3+} \rightleftharpoons BiOH^{2+} + H^+$; $\underline{I}$ = 3($NaClO_4$); [Bi^{3+}] determined by Bi-Hg electrode; main equilibrium is $6Bi^{3+} + H_2O \rightleftharpoons Bi_6(OH)_{12}^{6+} + 12H^+$, with $\log \underline{K}$ = 0.33	E3bg	O13
		Hydrolysis of Bi^{3+} gives $Bi_6O_6^{6+} + 12H^+$, with $-\log \underline{K}$ = 0.53 at 25° and $\underline{I}$ = 1($NaClO_4$), and at higher pH values $Bi_6O_6(OH)_3^{3+}$, with $\log \underline{K}$ = -8.1		T12
		Hydrolysis of $Bi_6O_6^{6+}$ (= $Bi_6(OH)_{12}^{6+}$) gives $Bi_9(OH)_{20}^{7+}$, $Bi_9(OH)_{21}^{6+}$ and $Bi_9(OH)_{22}^{5+}$; constants are listed		O14
1.61		$-\log \underline{K}$ for $Bi^{3+} + H_2O \rightleftharpoons BiOH^{2+} + H^+$; pH 0 - 2		D59
1.21		$-\log \underline{K}$ for $BiOH^{2+} + H_2O \rightleftharpoons Bi(OH)_2^{+} + H^+$		
0.58		$-\log \underline{K}$ for $6Bi(OH)_2^{+} \rightleftharpoons Bi_6(OH)_{12}^{6+}$		
12.36	25	$\log \underline{K}$ for $Bi^{3+} + OH^- \rightleftharpoons BiOH^{2+}$; $\underline{I}$ = 0.1	DISTRIB	B56
31.94		$\log \underline{K}$ for $Bi^{6+} + 3OH^- \rightleftharpoons Bi(OH)_3$		
32.90		$\log \underline{K}$ for $Bi^{3+} + 4OH^- \rightleftharpoons Bi(OH)_4^-$		
		Other measurements: D60		

18. Boranocarbonic acid, H_4BCO_2H					
≈8	10.24	25	pK_1, pK_2	E,g	M21
9.07	10.24	25	pK_1, pK_2 of 1,12-$B_{12}H_{12}(CO_2H)_2$, → 0	E,g	H23
19. Boric acid, H_3BO_3					
9.5078		0	Molal scale; equimolal concentrations (0.003 to 0.03 M)	Elch	M28
9.4374		5	of NaCl, borax and boric acid; extrapolated to $I = 0$		
9.3785		10	using extended Debye-Hückel equation		
9.3255		15			
9.2780		20			
9.2340		25			
9.1947		30			
9.1605		35			
9.1282		40			
9.1013		45			
9.0766		50			
9.0537		55			
9.0310		60			
			$pK = 2237.94/T + 0.016883T - 3.305$ (T in °K)		
			Thermodynamic quantities are derived from the results.		
9.440		5	Molal scale, I varied from 0.02 to 3 by adding NaCl;	Elch	O26
9.380		10	extrapolated to zero boric acid concentration at const-		
9.327		15	ant I, then to $I = 0$		
9.280		20			
9.237		25			
9.198		30			
9.164		35			
9.132		40			
9.080		50			
			Thermodynamic quantities are derived from the results.		

Name, Formula and p$\underline{K}$ value	T(^{o}C)	Remarks	Methods	Reference
9.380	10	$\underline{I}$ varied from 0.01 to 0.12; constants corrected using	Ela	O23
9.327	15	Debye-Hückel equation and extrapolated to $\underline{I}$ = 0		
9.280	20			
9.236	25			
9.197	30			
9.132	40			
9.080	50			
		p$\underline{K}$ = 9.023 + 8 × 10^{-5} $(76.7 - t)^2$ (t in ^{o}C)		
9.21	20	$\underline{I}$ = 0.04. The second p$\underline{K}$ of boric acid is greater than 14	E3ah	F46
8.98	25	$\underline{I}$ = 0.1 ($NaClO_4$)	E3bh	I4
9.00	25	$\underline{I}$ = 3($NaClO_4$)		
		At boric acid concentrations above 0.4 M, higher than		
		trimeric complexes are also formed		
9.00	25	$\underline{I}$ = 3($NaClO_4$); boric acid concentrations varied from 0.01	E3bh	I9
		to 0.60 M. Other equilibria were:		
		$3H_3BO_3 \rightleftharpoons H_4B_3O_7^- + H^+ + 2H_2O$, log $\underline{K}$ = –6.84,		
		$3H_3BO_3 \rightleftharpoons H_5B_3O_8^{2-} + 2H^+ + H_2O$, log $\underline{K}$ = – 15.44		
		Polymeric species are important at concentrations above		
		about 0.01 M		
		p$\underline{K}_a$ from 50-290°, 0.1 – 1.0 M in KCl		M64
		p$\underline{K}_a$ from 100° (8.93) to 350° (10.10)		S83
		p$\underline{K}_a$ from 100° (8.94) to 285° (9.70)		M40
		p$\underline{K}_a$ value in sea water		H24, M77
		p$\underline{K}_a$ from 0–50°, up to 1000 bars		W3
8.60	25	Allowing for polymeric species; $\underline{I}$ = 3($LiClO_4$)		M17
9.07	25	In D_2O; allowing for polymeric species; $\underline{I}$ = 3($LiClO_4$)		
		Other measurements: B102, B122, C55, E6, E8, F10, H10,		
		H29, I5, I6, K58, K72a, L19, L54, M56, O25, P62, S6, S23		

20. **Bromic acid**, $HBrO_3$				
1.02	25	In formamide	SOLY	D6
1.01	30			
0.91	35			
21. **Bromine**, Br_2				
8.48	10	$-\log \underline{K}$ for $Br_2 + H_2O \rightleftharpoons HBrO + H^+ + Br^-$	E,g	P67
7.92	25			
7.66	35			
7.49	50			
8.24	25			L37
8.16	25	Temperature-jump method	OTHER	E11
8.02	25	Corrected for tribromide formation	O	P47
8.05	25			D43
8.06	25		KIN	P31
22. (Aquo) **Cadmium ion**, Cd^{2+}				
10.2	25	$p\underline{K}$ for hydrolysis of Cd^{2+}; $\underline{I} = 3(NaClO_4 + Cd(ClO_4)_2)$; $c = 0.1$ to 1.45 $(Cd(ClO_4)_2)$; Cd_2OH^{3+} and $Cd_4(OH)_4^{4-}$ are also formed	E3bg, quin	B61
9.0	25	$p\underline{K}$ for hydrolysis of Cd^{2+}; $\underline{I} = 3(NaClO_4 + Cd(ClO_4)_2)$; $c = 0.01$ to 0.9 $(Cd(ClO_4)_2)$	E3bg	M29
9.49	100	$p\underline{K}$ for hydrolysis of Cd^{2+}; $c = 0.02$ $(CdCl_2)$	KIN	K108
9.3	25	$p\underline{K}$ for hydrolysis of Cd^{2+}	SOLY	G23
0.7	25	$p\underline{K}_b$ for $HCdO_2^- + H_2O \rightleftharpoons Cd(OH)_2 + OH^-$		
4.30	25	$\underline{I} = 3(NaClO_4)$, $p\underline{K}_b$ for $CdOH^+ \rightleftharpoons Cd^{2+} + OH^-$	DISTRIB	D73
3.44		$p\underline{K}_b$ for $Cd(OH)_2 \rightleftharpoons CdOH^+ + OH^-$		
2.58		$p\underline{K}_b$ for $Cd(OH)_3^- \rightleftharpoons Cd(OH)_2 + OH^-$		
1.72		$p\underline{K}_b$ for $Cd(OH)_4^{2-} \rightleftharpoons Cd(OH)_3^- + OH^-$ on assumption that $\log \underline{K}_n = 1/4 \log \underline{K}_1 \underline{K}_2 \underline{K}_3 \underline{K}_4 + (5 - 2n)/2)$		

Name, Formula and p$\underline{K}$ value	T(^{o}C)	Remarks	Methods	Reference
		log $(\underline{K}_n/\underline{K}_{n+1})$		
		log $\underline{K}$ for $Cd^{2+} + 4OH^- \rightleftharpoons Cd(OH)_4^{2-}$ is about 9.7 at 25°	POLAROG	L4
10.0	60	$-$log $\underline{K}$ for $Cd^{2+} \rightleftharpoons CdOH^+ + H^+$; $\underline{I}$ = 3(NaClO$_4$); c = 0.6-1.2 M	E3bg	B146
8.20		$-$log $\underline{K}$ for $2Cd^{2+} + H_2O \rightleftharpoons Cd_2OH^{3+} + H^+$		
10.3	25	$-$log $\underline{K}$ for $Cd^{2+} \rightleftharpoons CdOH^+ + H^+$; $\underline{I}$ = 3(NaClO$_4$)		M48a
9.13		$-$log $\underline{K}$ for $2Cd^{2+} + H_2O \rightleftharpoons Cd_2OH^{3+} + H^+$		
		Other measurements: B147, C14, G58, L52, S104		
23. Caesium hydroxide, CsOH				
		For alkalinity function for CsOH solutions, see L21, M89		
24. (Aquo) Calcium ion, Ca^{2+}				
1.02-1.14	0	p$\underline{K}_b$ for $CaOH^+$; m = 0.002– 0.02 $Ca(OH)_2$ in 0.003 – 0.01	Elb	B21
1.12-1.24	10	M $CaCl_2$ or 0.006 – 0.02 M KCl; values of p$\underline{K}_b$ depend on		
1.14-1.27	25	choice of γCl/γOH used to evaluate molality of hydroxylion.		
1.36-1.45	40			
1.34	15	p$\underline{K}_b$ for $CaOH^+$; $\underline{I}$ = 0.02 to 0.1 ($Ca(OH)_2$ + $CaCl_2$); $f\pm$ cal-	Elch	G40
1.37	25	culated assuming Davies' equation		
1.40	35			
1.37	0	p$\underline{K}_b$ for $CaOH^+$; $\underline{I}$ = 0.007 to 0.08	SOLY	B38
1.40	25	$\underline{I}$ = 0.02 to 0.08		
1.48	40	$\underline{I}$ = 0.04 to 0.10		
		extrapolated to $\underline{I}$ = 0 assuming Davies' equation;		
		$Ca(IO_3)_2$ in KOH solutions.		
1.30	25	p$\underline{K}_b$ for $CaOH^+$; $Ca(IO_3)_2$ in $Ca(OH)_2$ solutions; extrapolated using Davies' equation.	SOLY	D20
1.25-1.34	0	p$\underline{K}_b$ for $CaOH^+$; $\underline{I}$ = 0.18 to 0.30; value sensitive to choice of activity coefficient	KIN	B41
1.24-1.36	0	$\underline{I}$ = 0.025 to 0.08		

1.51	25	$\underline{I}$ = 0.03 to 0.15; extrapolated using Davies' equation; recalculation of data of G. Kilde, Z. Anorg. Allgem. Chem., 218 113 (1934)	SOLY	D16
1.31	25	Recalculation of data of F.M. Lea and G.E. Bessey, J. Chem. Soc., 1937 1612	C2	B39
10.3	75	pK_a for $Ca^{2+} = CaOH^+ + H^+$, in $Ca(NO_3)_2$	E3bg	C49
1.30	25	pK_b for $CaOH^+$; $c = 10^{-3}$M CaO		M41
2.80	60			
3.06	70			
3.50	80			
3.56	90			
3.88	98			
12.94		$\underline{I}$ = 0.05 - 3.0 (NaCl)	SOL	K68
13.04		$\underline{I}$ = 0.05 - 3.0 ($NaNO_3$)		
1.46	25	$\underline{I}$ = 0.13 to 0.24 ($Ca(OH)_2 + CaCl_2$); c = 0.02 – 0.03 $Ca(OH)_2$; extrapolated using Davies' equation	KIN	B39
1.29	25	$\underline{I}$ = 0.02 to 0.05; $f\pm$ calculated from Guggenheim's equation (Phil. Mag., 19, 588 (1935)	CAT,KIN	B40
1.03		Concentration constant; 0.2 – 1N $CaCl_2$; from salt effect on indicator	O3	K55
0.64	25	$\underline{I}$ = 3($NaClO_4$)	E2ah	C7
		For the acidity function of $Ca(OH)_2$ solutions from 0-95° and $\underline{I}$ = 0.01 to 0.20, see B22		
		Other measurements: G63		

25. (Aquo) Californium ion, Cf^{3+}

5.62	23	pK for hydrolysis of Cf^{3+}; $\underline{I}$ = 0.1 ($HClO_4$, $LiClO_4$)	DISTRIB	D46

26. Carbonic acid, H_2CO_3

6.577	0	Apparent pK values: double extrapolation procedure to	Elch	H34

Name, Formula and p$\underline{K}$ value	T(^{o}C)	Remarks	Methods	Reference
6.517	5	eliminate effect of added NaCl and to obtain values at		
6.465	10	zero bicarbonate concentration		
6.420	15			
6.382	20			
6.351	25			
6.327	30			
6.309	35			
6.296	40			
6.289	45			
6.287	50			
		$p\underline{K}_1 = 3404.71/\underline{T} - 14.8435 + 0.032786\underline{T}$ ($\underline{T}$ in $^{o}\underline{K}$)		
		Thermodynamic quantities are derived from the results.		
6.579	0	Apparent p$\underline{K}$ values; $\underline{I} = 0.004 - 0.2$, extrapolated to $\underline{I} = 0$	Elch	H36
6.517	5			
6.464	10			
6.419	15			
6.381	20			
6.352	25			
6.327	30			
6.309	35			
6.298	40			
6.290	45			
6.285	50			
6.514	5	Apparent pK values; $\underline{I} = 0.003 - 3$; extrapolated to $\underline{I} = 0$	E3bh	N11
6.421	15	by fitting to an extended Debye-Hückel equation		
6.349	25			
6.310	35			
6.294	45			
		$p\underline{K}_1 = 6.572 - 0.012173t + 0.00013329t^2$ (t in ^{o}C)		

6.583		0	Apparent p*K* values; 0.001 N in $KHCO_3$, KCl, HCl, and		Cl,Rld	S66
6.429		15	saturated CO_2 solutions			
6.366		25				
6.317		38				
6.35		25	Apparent p*K* value		E2b, quin	A62
6.35		25	Apparent p*K* value		Elc, quin	A61
6.38		25	Apparent p*K* values, molal scale,	1 atmosphere,	Cl	E17
5.90			*I* varied from 0.0001 to 0.1,	1035 atmosphere		
5.48				2050 atmosphere		
5.15				2930 atmosphere		
6.32		35		1 atmosphere		
5.85				1030 atmosphere		
5.45				2035 atmosphere		
5.12				2930 atmosphere		
6.32		45		1 atmosphere		
5.89				1015 atmosphere		
5.50				2010 atmosphere		
5.16				3000 atmosphere		
6.30		55		1 atmosphere		
5.86				1020 atmosphere		
5.49				2010 atmosphere		
5.17				2950 atmosphere		
6.31		65		1 atmosphere		
5.88				1050 atmosphere		
5.51				2060 atmosphere		
5.26				2800 atmosphere		
	10.625	0	*I* varied from 0.02 to 0.16; extrapolated to *I* = 0 using		E3ah	H44
	10.557	5	extended Debye-Hückel equation			
	10.490	10				
	10.430	15				

Name, Formula and pK value		T(°C)	Remarks	Methods	Reference
	10.377	20			
	10.329	25			
	10.290	30			
	10.250	35			
	10.220	40			
	10.195	45			
	10.172	50			
			$pK_2 = 2902.39/T - 6.4980 + 0.02379T$ (T in °K)		
			Thermodynamic quantities are derived from the results.		
	10.179	60	I varied from 0.005 to 0.1; extrapolated against I	E,g	C69
	10.153	70			
	10.142	80			
	10.140	90			
			$pK_2 = 2909.10T - 6.119 + 0.02272T$ (T in °K)		
	10.641	0	Double extrapolation, first to values in pure aqueous NaCl	E3ah	W13
	10.397	18	solutions, then against I to $I = 0$		
	10.32	25	$I = 0$		N10
6.35	10.33	25	$I = 1$(NaCl)	VAP	E16
6.29	10.17	50			
6.24	10.14	100			
6.33	10.25	150			
6.55	10.42	200			
6.42	10.13	100		Cl	R62
6.77	10.37	150			
7.27	10.80	200			
7.89	11.30	250			
8.70	12.0	300			
6.46	10.14	100		Cl	R61
6.81	10.41	156			

7.14		200			
	10.96	218			
6.34	10.25	25	*I* varied from 0.01 to 0.2; extrapolated against $I^{1/2}$	E3ag	M9
6.31	10.20	38	*I* varied from 0.01 to 0.2; extrapolated against $I^{1/2}$	E3ag	M10
3.81		5	True p*K* for $H_2CO_3 \rightleftharpoons H^+ + HCO_3^-$; high field conductivity measurements	C	W31
3.75		15			
3.76		25			
3.78		35			
3.80		38			
3.80		45			
3.68		0.5	True p*K* for H_2CO_3, calculated from apparent p*K*, using rates of hydration and dehydration		D49
3.88		25	True p*K* for H_2CO_3; from high field conductivity measurements, taking pK_{obs} = 6.352	C2	B46
3.75		23.5	True p*K* for H_2CO_3; from rapid-reaction measurements	KIN	S28
3.82		30.2			
3.89		35.6			
3.80		4	True p*K* for H_2CO_3; 150 atmospheres; pressure jump method.	C1	L50
6.364		25	*I* = 0, 1 bar	O1	R18
6.42		100			
6.76		150			
7.26		200			
7.79		250			
6.27		25	*I* = 0, 200 bar		
6.33		100			
6.66		150			
7.11		200			
7.62		250			
5.50		25	*I* = 0, 2000 bar		
5.61		100			
5.86		150			

Name, Formula and p*K* value	T(^{o}C)	Remarks	Methods	Reference
6.19	200			
6.52	250			
$pK_1 = 2416.1/T - 8.375 + 0.02225T$ (*T* in ^{o}K), over range 25-150°C				K96
10.32	25	*I* extrapolated to zero	E3bg	S9
		For p*K* of H_2CO_3 in sea water, see H25, M77		
		For values of pK_1 and pK_2 in D_2O/H_2O mixtures, see S9, S10		
		Other measurements: B110, B135, B159, C64, C65, D14, F7, F50, H47, K8, K13, K14, K15, K16, K17, K39, K59, M1, M52, M71, M95, N16, R49, S1, S108		
27. Caro's acid, see Peroxymonosulphuric acid				
28. (Aquo) Cerium (III) ion, Ce^{3+}				
8.1 16.3 26.0 32.8	rt	Cumulative constants for hydrolysis of Ce^{3+}; *I* = 1		K79
9.29		p*K* for hydrolysis of Ce^{3+}	E	S14
~9	25	p*K* for hydrolysis of Ce^{3+}; from hydrolysis of "pure" salts; *c* = 0.001 – 0.5 M $Ce_2(SO_4)_3$	E3ag	M87
		$Ce_3(OH)_5^{4+}$ was formed at 25° by hydrolysis of 0.05 M Ce^{3+} in 3 M $LiClO_4$		B66
		Other measurements: R10, S111		
29. (Aquo) Cerium (IV) ion, Ce^{4+}				
0.06	5	*I* = 1.1 – 4($HClO_4$, $NaClO_4$); *c* = 1 – 14 × 10^{-3} M; dimerization was important	O6	R27
–0.32	15			
–0.72	25			
–1.18	35			
0.70	25	*I* = 0.9 to 1.7($HClO_4$) *c* = 1 × 10^{-3} M Ce(IV); polymerisation was negligible.	O6	O3
–0.9 1.1	1.6	p*K* values for hydrolysis to $CeOH^{3+}$ and $Ce(OH)_2^{2+}$;	REDOX	B8

−1.15	0.82	25	$\underline{I} = 2(HClO_4, NaClO_4)$; $c = 3.5 \times 10^{-3}$ M Ce(IV); from pH-dependence of redox potential		
	−0.22	25	p$\underline{K}$ for $CeOH^{3+} \rightleftharpoons Ce(OH)_2^{2+} + H^+$; $HClO_4$ concentration from 0.2 – 0.4 M; from pH-dependence of redox potential	REDOX	S68
1.68		25	−log $\underline{K}$ for $2Ce^{4+} + 3H_2O \rightleftharpoons Ce_2(OH)_3^{5+} + 3H^+$; $\underline{I} = 3(NaNO_3)$; c = 0.01 – 0.10 $\underline{M}$ Ce^{4+}	E3bg, REDOX	D3
2.29			−log $\underline{K}$ for $2Ce^{4+} + 4H_2O \rightleftharpoons Ce_2(OH)_4^{4+} + 4H^+$		
1.98			−log $\underline{K}$ for $6Ce^{4+} + 12H_2O \rightleftharpoons Ce_6(OH)_{12}^{12+} + 12H^+$		
			Other measurements: D3		
30. Chloramine, see Monochloramine					
31. Chloric acid, $HClO_3$					
~−2.7			Theoretical prediction, based on structure		K77
32. Chlorosulphuric acid, $HClSO_3$			For p$\underline{K}$ in sulphuric acid, see B14		
~ −5.9			p$\underline{K}$ in H_2O estimated from relative strength in CF_3COOH	OTHER	B50
33. Chlorous acid, $HClO_2$					
1.94		25	Spectral differences extrapolated to zero time; c = 0.001 to 0.003 M $NaClO_2$, acidified with $HClO_4$; activity coefficients from Debye-Hückel equation	O6	L23
1.97		19-20	c = 0.001 – 0.1 M $NaClO_2$; extrapolation against $\underline{I}^{1/2}$	E3bg	D15
1.96		20			H82
1.99		23	"Practical" constant; concentration of $HClO_2$ ~0.25 M	E3bg,R2a	L47
			p$\underline{K}$ = −765/$\underline{T}$ + 3.912 ($\underline{T}$ in °K)	LIT	E8
			Other measurements: H85, L25, T1		
34. Chromic acid, H_2CrO_4					
	6.444	5		E3bg	L45
	6.478	15			

Name, Formula and pK values		T(°C)	Remarks	Methods	Reference
	6.488	25			
	6.524	35			
	6.569	45			
	6.642	60			
	6.472	15		O5	L45
	6.500	25			
	6.533	35			
	6.593	45			
	6.40	18	Titration of 0.025 M H_2CrO_4	E3bg	B115
	6.47	18	Titration of 0.04 M K_2CrO_4	E3bg	B122
	6.52	25	$\underline{I}$ = 0.0018 to 0.0028; $f\pm$ calculated from Davies' equation	E3bg	H92
	6.52	25	$\underline{I}$ = 0.002 to 0.004	O5	
	6.50	25	$\underline{I}$ = 0.01 to 0.16; extrapolated to $\underline{I}$ = 0; $\underline{K}$ = 0.023 for $Cr_2O_7^{2+} + H_2O \rightleftharpoons 2HCrO_4^-$	E3ag	N32
0.74		25	$\underline{I}$ about 0.16; HCl/KCl solutions	E3ch	
	6.60	20 ±2	$\underline{I}$ = 0.1; corrected to $\underline{I}$ = 0 by Davies' equation, tracer concentrations	DISTRIB	H13
–0.83		15	Concentration constants corrected for formation of	O6	T17
–0.61		25	CrO_3Cl^-; $\underline{I}$ = 1($LiClO_4$, LiCl, $HClO_4$, HCl)		
–0.42		35			
	6.49	25	$c = 2.5 \times 10^{-5}$ M	O5	B5
–0.98		25	In $HClO_4$ solutions		
0.51		25	In H_2SO_4 solutions, using H_- scale	O6	L18
0.76		25	In HCl solutions, using H_- scale	O6	
–1.91		25	In HNO_3 solutions, using H_0 scale	O6	
1.74		25	In H_3PO_4 solutions, using H_0 scale	O6	
–1.01		25	In $HClO_4$ solutions, using H_- scale	O6	
			$p\underline{K}_1$ varies with the proton source because of the formation of species such as $HCrO_3Cl$ and $HCrO_3(OSO_3H)$		

−0.81		25	*I* = 1.0; in HCl solutions, correcting for the formation of CrO_3Cl^-	O6	H12
			The equilibrium constant for $Cr_2O_7^{2-}$ + $H_2O \rightleftharpoons 2HCrO_4^-$ is 0.0265 at 20° and 0.0303 at 25°		D24
−0.18		20	*I* = 0.17–1.0($HClO_4$); extrapolated to *I* = 0	O1	L61
	6.5			E	I19
	6.68			POLAR	
	6.61			O1	
			Other measurements: B32, B134, G2, G3, G4, H98, J1a, J6, K76, M44, S16, S67, S70, S103, T16, T18.		

35. (Aquo) **Chromium(III) ion**, Cr^{3+}

3.66	25	p*K* for hydrolysis of Cr^{3+}; *I* = 0.0014 – 0.04; extrapolated to *I* = 0	E3ag	H46
3.95	25	p*K* for hydrolysis of Cr^{3+}; corrected for Cr^{3+} – SO_4^{2-} ion-pair formation; *I* = 0	C1	T27
3.34	46.2	p*K* for hydrolysis of Cr^{3+}; *I* = 0.068 ($LiClO_4$); from	O5	P58
3.01	63.6	variation of apparent stability constant of $CrNCS^{2+}$ with		
2.83	73.7	pH of $CrNCS^{2+}$ with pH		
2.65	84.8			
2.49	94.6			
4.05	15	*I* = 0.068 ($LiClO_4$); extrapolation from results at 46-95°		P58
3.82	25			
4.66	0	p*K* for hydrolysis of Cr^{3+}	C1	B78
4.01	25			
3.47	50			
2.99	75			
2.58	100			
4.26	20	p*K* for hydrolysis of Cr^{3+}; *I* = 0.5 ($NaNO_3$)	E3bg	J17
3.90	15	p*K* for hydrolysis of Cr^{3+}; *I* = 0	KIN	B128

Name, Formula and p*K* values	T(°C)	Remarks	Methods	Reference
4.13	25	p*K* for hydrolysis of Cr^{3+}; *I* = 0.5(Cr^{3+}, N^{2+}, ClO_4^-), molal scale, 1 bar	O1	S127
4.09		500 bars		
4.02		1000 bars		
3.99		1500 bars		
4.00		2000 bars		
3.96		2500 bars		
3.92		3000 bars		
10.50, 20.83, 30.06		cumulative constants for $Cr^{3+} + n\,OH^- \rightleftharpoons Cr(OH)_n^{(3-n)+}$	O	K72
4.1 ~5.6	20	Successive p*K* values for hydrolysis of Cr^{3+}; *I* = 0.1 ($NaClO_4$); rapid-flow measurements	E3ag	S49
4.10 5.55	25	Successive "practical" p*K* values for hydrolysis of Cr^{3+}; *I* = 0.04 to 0.4	E3ag	E27
3.96	25		O5	
		Other measurements B67, B79, B155, C18, D36, L5, S3.		
36. (Aquo) <u>Cobalt(II) ion</u>, Co^{2+}				
9.96	15	p*K* for hydrolysis of Co^{2+} at *I* = 0.25 and 0.75($NaClO_4$)	E3bg	B89
9.85	25			
9.62	35			
9.50	45			
8.9	30	p*K* for hydrolysis of Co^{2+}; *I* = 0.1(KCl)	E3bg	C14
8.7	100	p*K* for hydrolysis of Co^{2+}	KIN	K101
8.9	25	*I* = 0		S63
9.75	25	p*K* for hydrolysis of Co^{2+}; *I* = 3($BaClO_4$)		C42
9.44	25	$-\log K$ for $2Co^{2+} + 2H_2O \rightleftharpoons Co_2(OH)_2^{2+} + 2H^+$		
42.55	25	$-\log K$ for $6Co^{2+} + 6H_2O \rightleftharpoons Co_6(OH)_6^{6+} + 6H^+$ (Alternatively, $-\log \beta = 10.20$, $-\log \beta_{12} = 9.37$, $-\log \beta_{44} = 29.30$)		

		Other measurements: A7, D36, G21, P51.		
37. (Aquo) Cobalt(III) ion, Co^{3+}				
2.10	12.5	pK for hydrolysis of Co^{3+}; $I = 1(NaClO_4)$	O6	S123
1.98	18.5			
1.78	23.6			
1.71	28.2			
0.66	25	$I = 3(NaClO_4)$	KIN	C46
1.3	7	$I = 0.25$	KIN	H70
		The above values are uncertain because high cobaltic concentrations and low acidities favour formation of polynuclear species		
38. (Aquo) Copper(II) ion, Cu^{2+}				
		The pK for Cu^{2+} is not known; hydrolysis of Cu^{2+} gives almost entirely polynuclear complexes of the type, $Cu_n(OH)_{2n-2}^{2+}$; formation constants for $Cu_2(OH)_2^{2+}$ from 15–42° are given.		P34
8.0	25	pK for $Cu^{2+} \rightleftharpoons CuOH^+ + H^+$; $I = 3(NaClO_4)$: the major species formed is $Cu_2(OH)_2^{2+}$, with $-\log K = 10.6$	E3bg	B45
7.97	18	pK for $Cu^{2+} \rightleftharpoons CuOH^+ + H^+$; the major species formed is $Cu_2(OH)_2^{2+}$, with $-\log K = 10.89$	E3bg	P27
		Hydrolysis of Cu^{2+} gives $Cu_2(OH)_2^{2+}$, with $-\log K$ at 25° ranging from 10.5 to 10.9		H4
7.34	25	pK for $Cu^{2+} + H_2O \rightleftharpoons CuOH^+ + H^+$, corrected to $I = 0$	E3bg	A8
10.57		$-\log K$ for $2Cu^{2+} + 2H_2O \rightleftharpoons Cu_2(OH)_2^{2+} + 2H^+$		
10.78	20	$-\log K$ for $2Cu^{2+} + 2H_2O \rightleftharpoons Cu_2(OH)_2^{2+} + 2H^+$; $I = 0.1$ $(NaClO_4)$; rapid-flow	OTHER	W20
7.71	25	pK for $Cu^{2+} + H_2O \rightleftharpoons CuOH^+ + H^+$; $I = 0.1(KNO_3)$; $c = 0.2$–2.1 mM Cu^{2+}	E3bg	S132
10.99		$-\log K$ for $2Cu^{2+} + 2H_2O \rightleftharpoons Cu_2(OH)_2^{2+} + 2H^+$		

Name, Formula and pK values	T(°C)	Remarks	Methods	Reference
21.62		$-\log \underline{K}$ for $3Cu^{2+} + 4H_2O \rightleftharpoons Cu_3(OH)_4^{2+} + 4H^+$		
7.54	25	$p\underline{K}$ for $Cu^{2+} + H_2O \rightleftharpoons CuOH^+ + H^+$; $\underline{I} = 3(LiClO_4)$; $0.01-0.07$ M in Cu^{2+}	E3bg	O9
6.22		$-\log \underline{K}$ for $2Cu^{2+} + H_2O \rightleftharpoons Cu_2OH^{3+} + H^+$		
11.12		$-\log \underline{K}$ for $2Cu^{2+} + 2H_2O \rightleftharpoons Cu_2(OH)_2^{2+} + 2H^+$		
10.36		$-\log \underline{K}$ for $3Cu^{2+} + 2H_2O \rightleftharpoons Cu_3(OH)_2^{4+} + 2H^+$		
		Also gives results for dioxane-water solutions		
8.1	25	$p\underline{K}$ for $Cu^{2+} + H_2O \rightleftharpoons CuOH^+ + H^+$; $\underline{I} = 0.05$ $\underline{m}$ $(NaClO_4)$; ion-selective electrode	OTHER	P21
7.93		$\underline{I} = 0$		
16.4		$-\log \underline{K}$ for $Cu^{2+} + 2H_2O \rightleftharpoons Cu(OH)_2 + 2H^+$		
14	25	$-\log \underline{K}$ for $Cu^{2+} + 2H_2O \rightleftharpoons Cu(OH)_2 + 2H^+$; $\underline{I} = 0.001$; ion-selective electrode	OTHER	V21
7.22	25	$p\underline{K}$ for $Cu^{2+} + H_2O \rightleftharpoons CuOH^+ + H^+$; $\underline{I} = 3(NaClO_4)$	E3bg	K3
7.71		In D_2O; $\underline{I} = 3(NaClO_4)$; $c = 50-90$ mM Cu^{2+}		
10.75		$-\log \underline{K}$ for $2Cu^{2+} + 2H_2O \rightleftharpoons Cu_2(OH)_2^{2+} + 2H^+$; $\underline{I} = 3(NaClO_4)$		
11.46		In D_2O; $\underline{I} = 3(NaClO_4)$; $c = 50-90$ mM Cu^{2+}		
~10	25	Prediction of $p\underline{K}$ for $Cu(OH)_2 \rightleftharpoons HCuO_2^- + H^+$		M7
13.1	25	$p\underline{K}$ for $HCuO_2^- \rightleftharpoons H^+ + CuO_2^{2-}$	SOLY	
<12.8		$p\underline{K}$ for $Cu(OH)_2 \rightleftharpoons HCuO_2^- + H^+$	SOLY	J13
		Other measurements: A5, C14, C28, C58, D18, F51, K12, K108, M20, O6, Q3.		
39. (Aquo) Curium(III) ion, Cm^{3+}				
6.05	23	$p\underline{K}$ for hydrolysis of Cm^{3+}; $\underline{I} = 0.1(LiClO_4, HClO_4)$	DISTRIB	G75
5.92	23	$\underline{I} = 0.1(LiClO_4, HClO_4)$	DISTRIB	D46
7.83		$p\underline{K}$ for $CmOH^{2+} + H_2O \rightleftharpoons Cm(OH)_2^+ + H^+$; $\underline{I} = 0.1$		M102
		Other measurements: K74		

40. Cyanic acid, HCNO					
3.76		0	Corrected to I = 0 by extended Debye-Hückel equation; extrapolated to zero time to allow for decomposition	E3ag	C5
3.64		10			
3.57		18			
3.46		25			
3.37		35			
3.48		45			
3.54		18	I varied from 0.06 to 0.2; ionic strength correction doubtful; extrapolated to zero time	E3ag	J4
3.47		25	I = 0	E3ag	M114
3.70		20	I = 0	E3bg	B97
3.57		26			
3.69		33			
			pK = 350/T + 2.732 (T in ^{o}K)	LIT	E8
			Other measurements: A28, B98, T6		
41. Decahydrodecaboric acid, $H_2B_{10}H_{10}$			For acidity function, see M103		
42. Decavanadic acid, $H_6V_{10}O_{28}$					
3.65		20	pK_5; I = 0.2; also log K = − 7.50 for $10VO_2^+ + 8H_2O \rightleftharpoons H_2V_{10}O_{28}^{4-} + 14H^+$; and log K = − 23.5 for $2.5\ V_4O_{12}^{4-} + 3H_2O \rightleftharpoons HV_{10}O_{28}^{5-} + 5OH^-$	E3ag	S12
3.6	5.8	25	pK_5, pK_6; I = 1($NaClO_4$); total vanadium concentrations 2×10^{-3} – 2×10^{-2}M; also log K = −6.75 for $10VO_2^+ + 8H_2O \rightleftharpoons H_2V_{10}O_{28}^{4-} + 14H^+$	E3ag	R43
3.5		25	I = 3($NaClO_4$)	O5	
3.70			I = 1($NaClO_4$); 0.025 – 0.1 M in vanadate	E	C21
	6.12		I = 1($NaClO_4$); rapid titration	E	
4.45	7.52	20	I = 0.1(NMe_4Cl); rapid-reaction studies; complex formation occurs with alkali cations	O5	S51
2.19		33	Saturated Na_2SO_4 solution; up to 0.1 M Na_3VO_4 solutions; also log K = 2.4 for $H_2V_{10}O_{28}^{4-} + 14H^+ \rightleftharpoons 10VO_2^+ + 8H_2O$	CRYOSC	N17

Name, Formula and p$\underline{K}$ values	T(^{o}C)	Remarks	Methods	Reference
		Other measurements: C20		
43. Deuterium oxide, D_2O				
15.526	10	Molal scale; extrapolated to $\underline{I}$ = 0	Elch	C50, C56
15.136	20			
14.955	25			
14.784	30			
14.468	40			
14.182	50			
15.439	10	Molar scale		
15.049	20			
14.869	25			
14.699	30			
14.385	40			
14.103	50			
17.224	10	Mole fraction scale		
16.834	20			
16.653	25			
16.482	30			
16.166	40			
15.880	50			
		$p\underline{K}_m = 4913.14/\underline{T} - 7.5117 + 0.0200854\underline{T}$ ($\underline{T}$ in ^{o}K) Thermodynamic quantities are derived from the results.		
15.08	15	Molar scale; $\underline{I}$ = 0.04 to 0.10, extrapolated to $\underline{I}$ = 0	E2bh	W45
14.71	25			
14.37	35			
14.807	20	Molal scale, taking $p\underline{K}_w$ = 14.073 for H_2O	E3ah	S50
14.812	25	Molal scale	ANALYT	K35
14.81	25	Using 0.01 M $Ba(OH)_2$ in H_2O/D_2O mixtures; molal scale.	E3ag	S8

14.86	25	Molar scale, I = 0	E3bh	G47
14.856	25	Molar scale, I = 0	E3bh	G48
14.80	25	I = 0	Elcd	A2
		For pK values of H_2O/D_2O mixtures, see G47, G48		
14.84	25		E3ad	G50
14.86	25		E3bg	P29
		Also gives pK_w values for H_2O/D_2O mixtures		
14.957	298 °K	Molal scale, I = 0.06–0.09, extrapolated to I = 0	E2a	S74
14.184	323			
13.571	348			
13.115	373			
12.746	398			
12.459	423			
12.246	448			
12.096	473			
12.004	498			
11.968	523			
		$pK = -36319/T - 104.37 \ln T + 0.10287T + 2.61.10^6/T^2 + 671.15$		
14.845	25		E3bd	G50
		For pK_w values in mixed solvents, see G30		
44. Deutero-ammonia, ND_3				
9.757	20	I = 0; in D_2O; taking pK of NH_3 in H_2O as 9.265	E3ah	S50
4.9		pK_b; rough estimate		L32
45. Deutero-arsenic acid, D_3AsO_4				
2.596	25	pK_1 in D_2O; I = 0; from measurements in D_2O/H_2O mixtures	E3ag	S8
46. Deuterocarbonic acid, D_2CO_3				
11.332	5	pK_2, in D_2O	E3ad	P1

Name, Formula and pK values		T(°C)	Remarks	Methods	Reference
11.259		10			
11.192		15			
11.131		20			
11.070		25			
11.029		30			
10.984		35			
10.946		40			
10.912		45			
10.881		50			
6.77		25	Apparent pK in D_2O; taking pK_2 for CO_2 in H_2O = 6.35	E3a,quin	C64
	10.96	25	In D_2O; meter standardized in H_2O; pK_2 for H_2CO_3 in H_2O taken as 10.33	E3ag	G43
	10.93	25	In D_2O; taking pK_2 for H_2CO_3 in H_2O as 10.33	E3ag	C65
47. Deuterochloric acid, DCl					
			For Hammett acidity function in D_2O, see H78		
3.58		25	pK for solution in N,N-dimethyl formamide	O1	S78
48. Deuterocyanic acid, DCN					
8.97		23±2	pK in D_2O; I =0.11(KNO_3) (compare 9.03 for pK in H_2O)	E3bg	D67
49. Deuterodisulphuric acid, $D_2S_2O_7$					
			For pK_a in concentrated H_2SO_4, see F33		
50. Deuterohydrazine, N_2D_4					
	9.08	17	"Practical" constant; in D_2O; I = (KCl)	E3dg	F13
	8.69	30			
	9.11	18	"Practical" constant, in D_2O; I = 1(KCl)	E3dg	B136

51. <u>Deuterohydrazoic acid</u>, DN_3					
5.01		20	In D_2O; $I = 0$	E3bg	B143
5.115		25	In D_2O	E3bg	S7
52. <u>Deutero-iodic acid</u>, DIO_3					
1.15		25	In D_2O; taking pK of HIO_3 in H_2O = 0.85	C1,R1d	M6
53. <u>Deuterophosphoric acid</u>, D_2PO_4					
2.350		25	In D_2O; $I = 0$; extrapolated from measurements in D_2O/H_2O mixtures	E3ag	S8
2.362		25	In D_2O; taking pK_1 for H_3PO_4 in H_2O as 2.128	C	M6
2.31		25	In D_2O; meter standardised in H_2O; taking pK_1 for H_3PO_4 in H_2O as 2.11	E3ag	G43
	7.8846	5	In D_2O; using KD_2PO_4/Na_2DPO_4 mixtures from 0.005 to 0.025 M, and NaCl 0.005 M; extrapolated to $I = 0$	E1a	G14
	7.8499	10			
	7.8233	15			
	7.7986	20			
	7.7796	25			
	7.7667	30			
	7.7547	35			
	7.7484	40			
	7.7433	45			
	7.7435	50	$pK_2 = 2202.11/T - 5.9823 + 0.021388T$ (T in ^{o}K) Thermodynamic quantities are derived from the results.		
	7.750	25	In D_2O; extrapolated from measurements in D_2O/H_2O mixtures; taking pK_2 for H_2PO_4 in H_2O as 7.290; $I = 0$	E,quin	R51
2.188	7.666	20	In D_2O; $I = 0$; taking pK_1 and pK_2 for H_3PO_4 in water as 1.983, 7.207	E3ah	S50

Name, Formula and pK values		T(°C)	Remarks	Methods	Reference
54. Deuterosulphuric acid, D_2SO_4					
2.33		25	In D_2O; pK_2; $I = 0$, from plot of solubility of Ag_2SO_4 in dilute D_2SO_4 in D_2O as function of I from 0.1 to 1.0 M D_2SO_4	SOLY	L39
2.69		50			
2.98		75			
3.29		100			
3.59		125			
3.91		150			
4.21		175			
4.52		200			
4.84		225			
			$pK_2 = 0.0123175T - 22.86565/T - 1.253986$ (T in °K)		
			Thermodynamic quantities are derived from the results.		
			Other measurements of pK_2, see B7, D62		
			For Hammett acidity function in D_2O, see H78		
			For self-dissociation constant of D_2SO_4 see F33		
			For the D_o acidity function in D_2SO_4 solutions, see S76		
55. Diamidophosphoric acid, $(NH_2)_2PO_2H$					
1.051	5.299	0.0	$I = 0.5$, extrapolated to $I = 0$	E3bg	P23
1.098	5.103	8.0			
1.194	4.998	21.4			
1.279	4.889	30.0			
1.404	4.780	40.0			
	4.92	18	$I = 0$	C1	K40
	4.83	25			
56. Diamidothiophosphoric acid $(NH_2)_2POSH$					
2.0	4.3	20	$I = 0$	E3bg	P23

57. Difluorophosphoric acid, HPO_2F_2								
						For basic pK in sulphuric acid, see B22		
58. Diimidotriphosphoric acid, $H_2O_3P.NH.PO_2H.NH.PO_3H_2$								
~1	~2	3.03	6.61	9.84	25	Concentration constants, I = 0.1 (NMe_4Br)	E3bg	I13
~0	~2	3.83	7.02	9.92		I = 0.2 (NMe_4Br)		
~0	~2	3.94	7.74	9.95		I = 0.3 (NMe_4Br)		
~1	~2	3.36	6.86	10.00		I = 1 (NMe_4Br)		
						$f\pm$ assumed same as for HBr		
						Values of pK_1 and pK_2 could be seriously in error because of experimental limitations		
~1	~2.2	3.24	6.80	9.50	37	I = 0.1; same remarks as above	E3bg	I12
~1	~2.4	3.59	7.02	9.28	50	I = 0.1		
59. Diperosmic acid, See Osmic(VIII) acid								
60. Diperruthenic acid, (hydrated RuO_4)								
11.17					20?	Distribution between CCl_4 and water	DISTRIB	M38
14.24						pK_b for $H_2RuO_5 \rightleftharpoons HRuO_4^+ + OH_2^-$		
11.9								S77
61. μ-Disulphidohexaoxodiphosphoric acid, $S_2(PO_3H_2)_2$								
2.5	6.65				23		E3bg	N31
62. Disulphuric acid, $H_2S_2O_7$								
−12	−8					Theoretical predictions (Ricci's method) of pK_1 and pK_2		G37
−13	−8					Theoretical predictions (Pauling's method) of pK_1 and pK_2		
1.85					10	pK in concentrated H_2SO_4; molal scale	FP	B17
2.52					20	pK in concentrated H_2SO_4; molal scale		B105

Name, Formula and pK values	T(°C)	Remarks	Methods	Reference
63. Dithionic acid, $H_2S_2O_6$				
−3.4 −0.2		Theoretical predictions of pK_1 and pK_2 from structure		K77
0.49	25	pK_2; I = 2		K43
64. Dithionous acid, see Hyposulphurous acid				
65. Dodeca-antimonic acid, $H_{12}(Sb(OH)_6)_{12}$				
<1.55 <1.55 <1.55	25	pK_1, pK_2, pK_3	E3b	L21
1.55 2.95 4.35		pK_4, pK_5, pK_6		
5.75 7.15		pK_7, pK_8, all at I = 0.5 (NMe_4Cl); this acid exists in equilibrium with mononuclear antimony species at Sb(V) concentrations above 10^{-3} M		
66. Dodecahydrododecaboric acid, $H_2B_{12}H_{12}$		For acidity function, see M103		
67. Dodecanomolybdoceric acid				
4.16 7.15 9.13 11.25	20	pK_5, pK_6, pK_7, pK_8; I = 0.1		T5a
68. Dodecanomolybdouranic acid				
4.31 7.33 9.46 9.86	20	pK_5, pK_6, pK_7, pK_8; I = 0.1		T5a
69. Dodecatungstic acid, $H_{10}W_{12}O_{11}$				
~3.6 5.27 6.28	20	pK_8, pK_9, pK_{10}; I = 0.1(NaCl): rapid-reaction technique	E3ag	S52
70. (Aquo) Dysprosium(III) ion, Dy^{3+}				
8.10	25	pK_a for hydrolysis of Dy^{3+}; titration of 0.004-0.009 M $Dy(ClO_4)_3$ with 0.02 M $Ba(OH)_2$; I = 0.3 ($NaClO_4$)	E3b	F49

71. (Aquo) Einsteinium(III) ion, Es^{3+}				
		pK_1		H102
72. (Aquo) Erbium(III) ion, Er^{3+}				
7.99	25	pK_a for hydrolysis of Er^{3+}; titration of 0.004-0.009 M $Er(ClO_4)_3$ with 0.02 M $Ba(OH)_2$; $I = 0.3(NaClO_4)$	E3b	F49
9.0	25	pK for hydrolysis of Er^{3+}; c = 0.2-0.8M Er^{3+}; $I = 3(NaClO_4)$	E	B152
17.2	25	$-\log K$ for $Er^{3+} + 2H_2O \rightleftharpoons Er(OH)_2^{+} + 2H^+$; $I = 3(LiClO_4)$	E3bg	A26
13.72	25	$-\log K$ for $2Er^{3+} + 2H_2O \rightleftharpoons Er_2(OH)_2^{4+} + 2H^+$		
17.4	25	$-\log K$ for $Er^{3+} + 2D_2O \rightleftharpoons Er(OD)_2^{+} + 2D^+$; $I = 3(LiClO_4)$	E3bg	A27
14.29	25	$-\log K$ for $2Er^{3+} + 2D_2O \rightleftharpoons Er_2(OD)_2^{4+} + 2D^+$		
73. (Aquo) Europium(III) ion, Eu^{3+}				
8.31	25	pK_a for hydrolysis of Eu^{3+}; titration of 0.004-0.009 M $Ba(OH)_2$; $I = 0.3(NaClO_4)$	E3b	F49
~8.8	25	pK for hydrolysis of Eu^{3+}; hydrolysis of "pure" salt; c = 0.001-0.01 M $Eu_2(SO_4)_3$	E3ag	M87
8.03	25	$I = 0$	E3bg	U2
		Other measurements: M31		
74. (Aquo) Fermium(III) ion, Fm^{3+}				
3.8		$-\log K$ for $Fm^{3+} + H_2O \rightleftharpoons FmOH^{2+} + H^+$; $I = 0.1(LiClO_4)$; pH 4.3 - 5.09	DISTRIB	H102
75. Ferric ion, see Iron(III) ion				

Name, Formula and p$\underline{K}$ values		T(°C)	Remarks	Methods	Reference
76. **Ferricyanic acid**, $H_3Fe(CN)_6$					
<1		25	pK_3; $c \approx 10^{-3}$	E3bg	J16
77. **Ferrocyanic acid**, $H_4Fe(CN)_6$					
2.2	4.17	25	pK_3, pK_4; $c \approx 10^{-3}$; I = 0.01 to 0.5; extrapolated to I = 0	E3bg	J16
2.57	4.35	25	pK_3, pK_4; I = 0		H94
3	4.3	17	pK_3, pK_4; I = 0	E3bg	N27
	4.17	25	pK_4; I = 0.001 to 0.25; extrapolated against $I^{1/2}$	E3bg	L8
2.3	4.28	25	pK_3, pK_4 variation of redox potential with pH; extrapolated to I = 0.	REDOX	H21
	4.25	25	pK_4; variation of redox potential with pH Other measurements: K53	REDOX	K63
78. **Ferrous ion**, see Iron(II) ion					
79. **Fluoroberyllic acid**, H_2BeF_4					
2.46		26		E	G32
2.63			freezing-point lowering	OTHER	
80. **Fluoroboric acid**, $H_3O^+BF_4^-$					
-4.9			pK_a in water, based on relative strength in CF_3COOH		B50
81. **Fluorophosphoric acid**, H_2PO_3F					
≈0.5	4.80	40	"Practical" constants	E3b	D50
	5.12	25		R	R60
82. **Fluorosulphuric acid**, $HFSO_3$					
			For pK_a in H_2SO_4, see B14		

83. Fluorotriphosphoric acid, $H_4P_3O_9F$						
6.2			20	pK_4 No details		F16
84. (Aquo) Gadolinium(III) ion, Gd^{3+}						
≈8.8			25	pK for hydrolysis of Gd^{3+}; hydrolysis of "pure" salt; c = 0.001-0.01 M $Gd_2(SO_4)_3$	E3ag	M87
8.35			25	pK_a for hydrolysis of Gd^{3+}; titration of 0.004-0.009 M $Gd(ClO_4)_3$ with 0.02 M $Ba(OH)_2$; I = 0.3($NaClO_4$)	E3b	F49
8.20			25	I = 3($LiClO_4$)		A26
8.46				I = 0.1		M102
8.34			25	I = 3($LiClO_4$); D_2O solution		A27
8.27			25	I = 0	E3bg	U2
7.5			25	$-\log K$ for $Gd^{3+} + H_2O \rightleftharpoons GdOH^{2+} + H^+$; I = 0.5(H,Li)ClO_4	DISTRIB	G74
7.3			23	I = 1($NaClO_4$)	E3bg	K80a
				Also $-\log K$ = 14.6 for $Gd^{3+} + 2H_2O \rightleftharpoons Gd(OH)_2^+ + 2H^+$, $-\log K$ = 21.9 for $Gd^{3+} + 3H_2O \rightleftharpoons Gd(OH)_3 + 3H^+$, and $-\log K$ = 19.0 for $3Gd^{3+} + 4H_2O \rightleftharpoons Gd_3(OH)_4^{5+} + 4H^+$		
				Other measurements: D8, N37		
85. (Aquo) Gallium(III) ion, Ga^{3+}						
3.34-3.43			25	pK for hydrolysis of Ga^{3+}; c = 0.004-0.25 M in Ga^{3+}	E3bg	M88
2.95			25	pK for hydrolysis of Ga^{3+}; I = 0.5($NaClO_4$)	O6	W27
2.8	3.5		18	Successive pK values for hydrolysis of Ga^{3+}	E3ah	F47
10.3	11.7		18	Successive pK values for hydrolysis of $Ga(OH)_4^-$ to $Ga(OH)_5^{2-}$ and $Ga(OH)_6^{3-}$		
6.85			20	pK for $Ga(OH)_3 + H_2O \rightleftharpoons Ga(OH)_4^- + H^+$; c = 0.005-0.025 M in $Ga(OH)_4^-$; from pH of hydrolysed alkali-metal salts		I28
1.40	1.75	2.12	20	Successive pK values for hydrolysis of Ga^{3+} to $GaOH^{2+}$, $Ga(OH)_2^+$ and $Ga(OH)_3$; I = 1(NaCl)	DISTRIB	A24

Name, Formula and pK values			T(°C)	Remarks	Methods	Reference
2.91	3.70	4.42	25	Successive pK values for hydrolysis of Ga^{3+}; $I = 0.1$	O8	N18
2.92	3.77	4.75	25	$I = 0.1=1.0$, extrapolated to $I = 0$	O8	B71
0.44	0.80	1.05	20	Successive pK values for hydrolysis of Ga^{3+}	DISTRIB	S22
				Polymeric species are formed when Ga^{3+} is hydrolysed		G6
86. Germanic acid, H_4GeO_4 (H_2GeO_3)						
9.02	12.82		25	pK_1; pK_2; $I = 0.5(NaClO_4)$	E3bg	H1
8.98	11.74		25	$I = 1(NaClO_4)$		
				Below 0.004 M, germanic acid is mainly monomeric; at higher concentrations an octagermanic anion is also formed		
9.03	12.33		25	pK_1, pK_2; $I = 0.5(NaCl)$; log $K = 29.14$ for $8Ge(OH)_4 + 3OH^- \rightleftharpoons (Ge(OH)_4)_8(OH)_3^{3-}$	E3bg,h	I7
8.98			15	$I = 0$	E3cg	A36
8.92			20			
8.73			25			
8.62			30			
9.1	12.7		12	$I = 2(KCl)$	E3b	C9
9.08				In 0.5 M Na_2SO_4	E	L55
	12.43		25	$I = 3(NaCl)$	E3bh	I10
	12.31		32	In saturated Na_2SO_4 solution	CRYOSC	K92
				Other measurements: G24, G80, N23, P66, R46, S45		
	$pK_1 = 1093/T + 4.808$ (T in °K)				LIT	E8
87. Gold(III) hydroxide, $Au(OH)_3$						
<11.7	13.36	>15.3	25	Successive pK values for ionization to $H_2AuO_3^{2-}$, $HAuO_3^{2-}$ and AuO_2^{3-}	SOLY	J13

88. (Aquo) Hafnium(IV) ion, Hf^{4+}							
−0.12	0.23	0.42	0.52	25	Successive pK values for hydrolysis of Hf^{4+}; I = 1($HClO_4$); using low (radio-isotope) Hf^{4+} concentrations; at concentrations above 10^{-3} M polymers (mainly trimers and tetramers) are also formed	DISTRIB	P39
0.15				25	pK for Hf^{4+} $H_2O \rightleftharpoons HfOH^{3+} + H^+$	DISTRIB	N43
1.10				25	pK for $Hf^{4+} + H_2O \rightleftharpoons HfOH^{3+} + H^+$; competitive complex function by F^-; I = 4(Na^+,H^+)ClO_4	OTHER	
−0.04				25	I = 0.1-1(KNO_3)		N20
89. Heptamolybdic acid, $H_6Mo_7O_{24}$							
~3.7	4.33			25	pK_5 , pK_6; I = 3($NaClO_4$); concentration constants; also log K = 57.7 for $7MoO_4^{2-} + 8H^+ \rightleftharpoons Mo_7O_{24}^{6-} + 4H_2O$	Elcg,H	S17
90. Hexadecapolyphosphoric acid, $H_{18}P_{16}O_{49}$							
~2	2.92	6.48	6.08	25	Concentration constants: I = 1(NMe_4Br); $f\pm$ assumed same	E3bg	I12
~2	2.64	6.48	8.02	37	as for HBr; first two of these pK values may be seriously		
~2	2.52	6.50	8.18	50	in error because of experimental difficulties		
91. Hexafluorosilicic acid, H_2SiF_6							
1.92					pK_2 No details		S119
92. Hexametaphosphoric acid, $H_6P_6O_{18}$							
2	5.60	7.82			No details		K10
93. Hexaminotriphosphazene, $N_3P_3(NH_2)_6$							
<3.2	7.65			25	pK_1, pK_2; I = 0	E	F11
	7.70			25	pK_2; I = 0	E	F12

Name, Formula and p$\underline{K}$ values				T(^{o}C)	Remarks	Methods	Reference
94. Hexapolyphosphoric acid, $H_8P_6O_{19}$							
~2.1	2.19	5.98	8.13	25	Concentration constants; $\underline{I}$ = 1(NMe_4Br); $f\pm$ assumed same as for HBr; lowest two p$\underline{K}$ values uncertain because of experimental difficulty	E3bg	I13
~1.3	2.22	5.83	8.02	37	Concentration constants, as above	E3bg	I12
~1.3	2.22	5.81	8.00	50			
95. (Aquo) Holmium(III) ion, Ho^{3+}							
8.04				25	pK_a for hydrolysis of Ho^{3+}; titration of 0.004-0.009 M $Ho(ClO_4)_3$ with 0.02 M $Ba(OH)_2$; $\underline{I}$ = 0.3($NaClO_4$)	E3b	F49
96. Hydrazine, N_2H_4							
8.26				15	$\underline{I} \rightarrow 0$	E,g	S4
8.14				20			
7.97				25			
7.75				30			
–0.88	8.11			20	$\underline{I}$ = 0	O2,E	S47
0.27	7.94			25	$\underline{I}$ = 0	E3bg	Y22
	8.24			15	$\underline{I}$ from 0.01 to 0.15; extrapolated to $\underline{I}$ = 0	E3dg	W4
	7.99			25			
	7.82			35			
	8.60			10	"Practical" constant; $\underline{I}$ = 1(KCl)	E3dg	B136
	8.40			18			
	8.20			25			
	8.15			25	"Practical" constant; $\underline{I}$ = 0.3($NaClO_4$)	E3bg	J5
	8.07			30	"Practical" constant; 0.02-0.05 M hydrazine	E3dg	H72
					Other measurements: B109, G34, H98		
					For H– acidity function of hydrazine see D39, F9, S24, S112a		

97. Hydrazinosulphuric acid, $^{+}NH_3NHSO_3^{-}$

3.85		0.0075 M solution	E3bg	A54

98. Hydrazoic acid, HN_3

4.72	25	*I* = 0	E1cg	Y23
4.65, 4.68	25	*I* = 0.01 to 0.03	E3bg	
4.62	22	*I* = 0.03 to 1.0; p*K* corrected using Debye-Hückel equation and extrapolated against *I*	O5	B156
4.59	25		E3ag	H98
4.68	20	*I* = 0	E3bg	B143
4.692	15	*I* = 0.02	E3cg	S55
4.686	17.5			
4.684	20			
4.682	22.5			
4.680	25			
4.680	27.5			
4.680	30			
4.44	25	*I* = 1($NaClO_4$)	E3bg	M19
4.78		*I* = 3($NaClo_4$)		
4.70	20	*I* = 0.01 to 0.04, extrapolated to *I* = 0	E3bg	B98
4.64	26			
4.58	33			
4.55	20	*I* = 0.1 to 1.3(KCl); extrapolated against $I^{1/2}$	E3b	Q4
−6.21	25	p*K* of monocation, $H_2N_3^{+}$, using H_0 function for H_2SO_4	DISTRIB	B6
−10.1		p*K* of dication, $H_3N_3^{2+}$, using H_0 function for H_2SO_4 and data by A. Hantzsch (*Ber.*, **63B** 1782 (1930)		

Other measurements: B122, H26, H79, O16, W22

99. Hydriodic acid, HI

−8.56	25		VAP	M30

Name, Formula and pK values	T(°C)	Remarks	Methods	Reference
~−9	25	Using Raoult's law	VAP	B34
~−9.5	25	Calculation from thermodynamic data		M3
		Other measurements: B50, E29		
100. Hydrobromic acid, HBr				
−8.34	25	pK, molal scale	VAP	O1
−8.69	25	from thermodynamic results of M.V. Ionin, Trans. Gorkovsk. Politekh. Inst., 25 35 (1969)	OTHER	E30
−8.60	25	molal scale	VAP	M30
−8.76	25	molal scale	VAP	
−8.85	20	1-20 m HBr, molal scale	NMR	W38
−7.73	54.8			
−9.01	6			S91
−9.60	30			
−10.73	60			
~−8	25	Using Raoult's law	VAP	B34
~−9	25	Calculation from thermodynamic data		M3
		For Hammett acidity function of HBr, see P11, V8		
		Other measurements: B50, E29		
101. Hydrochloric acid, HCl				
−7.3	0		VAP	R28
−6.8	10			
−6.4	20			
−6.1	25			
−5.9	30			
−5.4	40			
−5.1	50			

~−7.4	0		VAP	W43
~−7	25	Assuming free HCl is like free HCN		B34
−7	25	Calculation from thermodynamic data		
~−7	25	Calculation from thermodynamic data		M3
~−7		Assuming solubilities of free HCl and RCl (where R = CH_3, C_2H_5, etc.) in water fall in regular sequence		E2
~−6		Assuming $\underline{K}_{HF}/\underline{K}_{HCl} \approx \underline{K}_{H_2O}/\underline{K}_{H_2S} \approx 10^{-9}$		S41
−6.18	25	From partial HCl vapour pressure of 1.3-13 M HCl in 15.5 m LiCl	VAP	R29
−6.31		From partial HCl vapour pressure of 1.3-13 m HCl in saturated LiCl solution		
−6.19		From partial HCl vapour pressure of 12-17 m HCl in absence of LiCl		
−6.07	25	molal scale	VAP	M30
3.26	360	I = 0; in superheated steam, density 0.525 g/ml	Cl	P24
3.42	373	0.525		
3.47	378	0.525		
4.11	370	0.447		
4.14	373	0.447		
4.24	378	0.447		
4.32	383	0.447		
4.61	373	0.399		
4.74	378	0.399		
3.57	25	pK for solution in N,N-dimethyl formamide	O1	S78
~−10		pK of H_2Cl^+; theoretical prediction		S48
~−3		pK of HCl; theoretical prediction		

For pK values of HCl in superheated steam between 400 and 700°, with densities from 0.3 to 0.8 g/cm³, see F40

For pK value in absolute ethanol, see S41

For Hammett acidity function of HCl see B36, B106, D33

Name, Formula and p$\underline{K}$ values	T(^{o}C)	Remarks	Methods	Reference
		(in the presence of LiCl and NaCl), G27 and G28 (temperature range), P10 (in the presence of added salts), P20, V17		
		For H_A acidity function, see Y10		
		For H_- acidity function, see P46		
		For H_0', H_0''', H_R and H_R' acidity functions of HCl, see A51		
		Other measurements: B50, E29, T4		
102. Hydrocyanic acid, HCN				
9.216	25	Taking p$\underline{K}$ of *m*-bromophenol as 9.004; 0.01-0.05 M borax buffers; extrapolation to $\underline{I}$ = 0, using extended Debye-Hückel equation; freshly prepared cyanide solutions	O2	A33
9.63	10	$\underline{I}$ = 0.002 to 0.024; extrapolation to $\underline{I}$ = 0 using extended Debye-Hückel equation, freshly prepared cyanide solutions	E3bg	I30
9.49	15			
9.36	20			
9.21	25			
9.11	30			
8.99	35			
8.88	40			
8.78	45			
		Thermodynamiç quantitits are derived from the results		
9.36	20	$\underline{I}$ = 0.01 to 0.04, extrapolated to $\underline{I}$ = 0	E3bg	B98
9.19	26			
9.05	33			
9.30	28		E3b	G19
		$pK_1 = 2382/T + 0.568$, $\underline{T}$ in ^{o}K	LIT	E8
		Other measurements: A30, B118, B121, B122, G15, H33, K52		

103. Hydrofluoric acid, HF (H_2F_2)

3.18	25	for I = 0; 0.01-0.1 M in HF, 0.002-0.01 M in KF; over these temperatures, K_1 for $F^- + HF \rightleftharpoons HF_2^-$ is 3.4, 4.0, 4.7, 4.8, 4.9, 5.7, 5.8, 8, respectively	Cl,Rlb	E18
3.40	50			
3.64	75			
3.85	100			
4.09	125			
4.34	150			
4.58	175			
4.89	200			
		Data fit $pK = 2.75 + 295/T - 1.91 \log T + 0.014\, T$ (T in ^oK) Thermodynamic quantities are derived from the results		
3.21	25	for I = 0; 0.001 M in NaF; dilute HF solutions	Cl	E32
3.10	15	for I = 0; using Pb-Hg/PbF_2 instead of Ag/AgCl, 0.001-1.0 M in HF, K_1 for $F^- + HF \rightleftharpoons HF_2^-$ was 3.94, 3.86, 4.32 at 15, 25, 35°	Elch	B124
3.17	25			
3.25	35			
3.164	25	pK_1; I = 0.01-0.5, extrapolated to I = 0; ion-selective electrode, log K_2 = 0.7	OTHER	B28
3.189	25	I = 0.05-0.50, extrapolated to I = 0; $[F^-] = 10^{-5}$–$10^{-3}M$, ion-selective electrode	OTHER	V2
3.23	25	I = 0.008–0.05, extrapolated to I = 0; molal scale	E3c	P17
3.32	25	I = 3(KCl); fluoride-ion electrode	OTHER	N29
4.04		log K for $HF + F^- \rightleftharpoons HF_2^-$; I = 3(KCl)		
3.16	22	I = 0.0005–0.005; fluoride-ion electrode	OTHER	W5
4.3	25	log K for $HF + F^- \rightleftharpoons HF_2^-$		
		log K_d for $2HF \rightleftharpoons (HF)_2$ = 2.7		
2.96	0	Recalculation of data by E. Deussen (Z. Anorg. Allgem. Chem., **44** 312 (1905); K_1 for $F^- + HF \rightleftharpoons HF_2^-$ was 2.43, 2.70 at 0, 25°	C	W37
3.16	25			
3.16	25	Taking $a_{HF_2^-}/(a_{HF} \cdot a_{F^-})$ = 5.4	E	B129
~–9		pK of H_2F^+; theoretical prediction		S48

Name, Formula and pK- values	T(^{o}C)	Remarks	Methods	Reference
-12.50	0	log of autoprotolysis constant; I = 0.1, $KSbF_6$ For Hammett acidity function of HF, see B35, H103, M55, N6 (in ethanol-water mixtures), P20, V1. Other measurements: A20, A56, B29, B96, B132, B133, C33, C44, C45, D21, D64, F4, F44, R44, R45, R63(at 100, 156, 218°), S122	E	G82
104. Hydrogen peroxide, H_2O_2				
11.86	15	I = 0.05 to 4.8($NaClO_4$); extrapolated against $I^{1/2}$; c = 0.55 M H_2O_2	E3ag	E34
11.75	20			
11.65	25			
11.55	30			
11.45	35			
11.81	20		E3bg	K11
11.92	10		CALOR	S5
11.62	25			
11.34	35			
11.21	50			
12.11	0		KIN	J24
12.23	0		DISTRIB	
12.19	0		C1	
11.85	19		O1	J19
11.58	30	I = 0.1 (phosphate buffers) corrected to I = 0 by Debye-Hückel equation		M104
		In strong hydrogen peroxide solutions (above several per cent H_2O_2 in water), superacidity is observed, giving lower values of pK which pass through a flat minimum (8.7) near 50%		M81, K46
		Other measurements: E36		

105. Hydrogen polysulphide					
3.8	6.3	20	For H_2S_4; I = 0.1 ($NaClO_4$); rapid-flow measurements; 3.4 and 5.6 for H_2S_5	E3ag	S49
106. Hydrogen selenide, H_2Se					
	15.0	22	Estimated uncertainty ± 0.6 pH units; the direct titration of H_2Se with KOH gives low pK_2 values because of aerial oxidation	SOLY	W35
	14	25	Value needed to fit experimental $E_{1/2}$/pH plot	POLAROG	L44
3.89	11.0	25	$I \sim 0.03$; titration of H_2Se in the dark	E3bg	H9
3.73		25.9	c = 0.008-0.1 M H_2Se	C1	H75
3.77					B138
107. Hydrogen sesquioxide, H_2O_3					
9-10			Theoretical prediction		C70
108. Hydrogen sulphide, H_2S					
7.33		5		C1	W40
7.24		10			
7.13		15			
7.05		20			
6.97		25			
6.90		30			
6.79		40			
6.69		50			
6.62		60			
7.57		0	c = 0.001 to 0.017 M in H_2S	C1,R1b	L57
7.06		25			
6.82		50			
7.02		25	I varied from 0.01 to 0.17, phosphate buffers	O1	E22
	14.0	20	Extrapolation of measured pK versus alkali concentration	O7	

Name, Formula and p$\underline{K}$ values		$T(^{o}C)$	Remarks	Methods	Reference
7.0	14.7	0		FP,C	J3a
7.0	13.85	30		O5	M107
7.05	13.8	25		O	E23
6.5	12.1	90			
			$p\underline{K}_1 = 3760/\underline{T} - 55.06 + 20 \log \underline{T}$ ($\underline{T}$ in ^{o}K)		
		$25\text{-}150^{o}C$	$p\underline{K}_1 = 2694.9/\underline{T} - 7.43 + 0.01779\ \underline{T}$ ($\underline{T}$ in ^{o}K)		K95
		25-150	$p\underline{K}_1 = 2718.3/\underline{T} - 7.715 + 0.0184\ \underline{T}$ ($\underline{T}$ in ^{o}K)		K96
			$p\underline{K}_1 = 1132/\underline{T} + 3.191$ ($\underline{T}$ in ^{o}K)		E8
			$p\underline{K}_2 = 2793/\underline{T} + 4.590$		
		25-150	$p\underline{K}_2 = 2892.91/\underline{T} + 4.448 - 0.002514\ \underline{T}$		K97
7.26		10	Values of $p\underline{K}_2$ (obtained from titrations) given in this reference are probably too low	E3bg	T29
7.07		25			
6.99		35			
6.91		50			
6.96		18	c = 0.001-0.04 M in H_2S	E3ag,VAP	G54
6.87		20			
6.79		25			
6.66		35			
6.54		45			
6.91		25		E3bg	Y24
6.81		25	1 atmosphere pressure	C1	E20
6.68		25	500 atmosphere pressure		
6.56		25	1000 atmosphere pressure		
6.45		25	1500 atmosphere pressure		
6.37		25	2000 atmosphere pressure		
	14.75	0	Calculated from thermodynamic data and potential measurements		M33
	13.90	25			

6.88	14.15	20	I = 1(KCl); Hg electrode versus calomel	E	W25
6.99	12.89	25	Calculated from published thermodynamic data		P53
6.81	12.24	40			
6.54	10.68	80			
6.52		90			
6.59	9.27	120			
	8.55	138			
			Other measurements:A23, A55, E21, E31, K45, K72b, K99, K110, P18, R12, S48, S99, S110, S112, S115, T9, W1, W6, Z1		
109. Hydrogen telluride, H_2Te					
2.64		18	*c* = 0.003–0.09 M H_2Te	C1	H75
2					B138
	11	25	Value of pK_2 needed to fit $E_{1/4}$/pH plot	POLAROG	L44
	12.16	25	Value of pK_2 needed to fit $E_{1/4}$/pH plot	POLAROG	P12
110. Hydroperoxy radical, HO_2					
4.4		23	pK for $HO_2 \rightleftharpoons H^+ + O_2^-$; from pH-dependence of reaction with tetranitromethane; species generated by electron irradiation	KIN	C70
4.45		23	pH- dependence of rate of reaction with tetranitromethane	O	R2
4.5			pulsed radiolysis experiments	O	C71
~2			estimate		U1
~6		20	estimate		W16
111. Hydrosulphuric acid, see Hydrogen sulphide					
112. Hydroxylamine, NH_2OH					
6.186		15	I = 0.25, 1, 2.25 ($NaClO_4$); extrapolated to I = 0	E3bg	L62

Name, Formula and pK values	T(^{o}C)	Remarks	Methods	Reference
6.063	20	using Debye Hückel equation		
5.948	25			
5.730	35			
		$pK = 2775.7/T - 5.8899 + 0.0084782T$ (T in ^{o}K)		
		Thermodynamic quantities are calculated from the results		
6.04	20	for $I = 0$; taking pK of 3,4-dinitrophenol as 5.46, 5.42,	O2	R30
5.96	25	and 5.38		
5.84	30			
5.98	25	For $I = 0$	E3ag	H8
5.93	25		DISTRIB	
5.97	25	$I = 0.0023$ to 0.023; extrapolated against $I^{1/4}$	E3bg	B76
6.04	30	"Practical" constant; $I = 1(KCl)$	E3ag	B137
6.49	30	"Practical" constant; $I = 1(KCl)$; in D_2O	E3ag	
		Other measurements: F2, F43, I16, M39, M105, M106, R38, S134, W28		
113. Hydroxylamine-N,N-disulphonic acid, $HON(HSO_3)_2$				
11.85	25	pK_3; $I = 1.6(K_2SO_4?)$	E3ag	A18
114. Hydroxylamine-N-sulphonic acid, $HO.NH.OSO_2H$				
~12.5	Room	pK_2; $I = 1.5(K_2SO_4)$	E3ag	A18
12.38	64.2	$I = 1.6(Na_2CO_3$ or $Na_2SO_4)$		
12.20	73.8			
12.10	83.5			
115. Hydroxylamine O-sulphonate, $^{+}NH_3OSO_3^{-}$				
1.48	45	$I = 1(NaClO_4)$	E3ag	C3

116. Hydroxyl radical, ·OH				
11.9	23	Pulse radiolytic method		R1
11.8	~23	Pulse radiolysis; p$\underline{K}$ obtained from pH-dependence of rate of formation of radical ion, $\cdot CO_3^-$		W14
117. Hypobromous acid, HOBr				
8.91	10		O	F30
8.66	25			
8.49	25			
8.23	50			
8.80	15.65		E3bg	K21
8.60	25.28			
8.47	35.55			
8.36	45.55			
8.68	22	$\underline{I}$ = 0.02 to 0.1	E3bg	S71
8.69	20	c = 0.01-0.02 M BrO^-	E3bg	S72
		Other measurements: C18, F3, K27, L38, S86, S101		
118. Hypochlorous acid, HOCl				
7.825	0	Measured relative to p$\underline{K}_2$ of H_3PO_4; $\underline{I}$ = 0.05 to 0.2; extrapolated to $\underline{I}$ = 0	O3	M97
7.754	5			
7.690	10			
7.633	15			
7.582	20			
7.537	25			
7.497	30			
7.463	35			
7.49	10	For $\underline{I}$ = 0; c = 0.01 M HOCl	E3bg	F28
7.30	25			
7.18	35			

Name, Formula and pK values	T(°C)	Remarks	Methods	Reference
7.05	50			
7.50	10	For $\underline{I}$ = 0; c = 0.003 M HOCl	O5	O21
7.31	25			
7.19	35			
7.06	50			
7.52	25		E,O	P7
7.34	25		E	S114
7.13	40			
6.82	60			
		$p\underline{K} = 1.39.10^3/\underline{T} + 2.48$ ($\underline{T}$ in °K)		
7.82	0	Extrapolated to zero time, and to $\underline{I}$ = 0 using Debye-Hückel equation	E3ag	C4
7.72	10			
7.65	15			
7.53	25			
7.49	35			
7.46	45			
7.53	25		E3bg	H7
7.50	20		E3bg	S72
7.49	25			M113
7.66	0.6	For $\underline{I}$ = 0; using Debye-Hückel equation	E3bg	A53
7.55	20			
7.42	27	"Practical" constant; c = 0.25 M HOCl	E3bg, R2a	L47
		Other measurements: B119, D14, G5, G41, H81, H93, I2, K28, P8, S11, S73, S85, S86, S98, Y18		
119. Hypoiodous acid, HOI				
10.64	25	Also $p\underline{K}$ = 14.48 for $I_2OH^- \rightleftharpoons I_2O^{2-} + H^+$	E,h	C26
9.7	22		E3bg	J21
~11	25			F55

12.4				20		KIN	S84
9.49				25	pK_b for $HOI \rightleftharpoons I^+ + OH^-$; iodine electrode	E	M110
1.35				25	pK of H_2OI^+	O5	A25
1.54				25	pK of H_2OI^+; cells of type $Pt, I_2, Ag^+, H^+/Sat.KNO_3/I^-, I_2, H^+, Pt$	E	B37
120. Hyponitrous acid, $H_2N_2O_2$							
7.51				0	I ⩽ 0.06; from rates of decomposition	KIN	P55
7.22				20			
7.09				30			
	11.35			25			
	11.09			50			
	10.97			55			
	11.1			18		E3ag	
7.32				15	For I = 0	KIN	H99
7.21	11.54			25			
7.17				35			
6.92	10.90			45	Using borate buffers in determining K_2		
	11.10			45	Using NaOH solutions		
7.05				25	For I = 0	E3bg	L13
	11.4			25		O3	
6.75	10.85			25	I = 1	KIN	B140
					Other measurements: A3, P54		
121. Hypophosphoric acid, $H_4P_2O_6$							
	2.19	6.77	9.48	20	"Practical" constants, I = 0.1 (KCl)	E3bh	S54
<2	2.81	7.27	10.03		Concentration constants; titration of 0.01 M $Na_4P_2O_6$ in 0.049 M HCl with 0.1 M NaOH	E3bh	T23

Name, Formula and pK values				T(°C)	Remarks	Methods	Reference
122. Hypophosphorous acid, H_3PO_2							
1.23				25	For I = 0	C1,R1c	P16
1.07				18	"Practical" constant; titration of 0.11 N H_3PO_2 with 0.11 N NaOH	E3bg	M99
1.02				16	I = 0.16	E3a	G66
1.12				30	I = 0.16		
1.2				45	I = 0.57		
1.03				16	I = 1.13(KCl); concentration constant		
1.31				25	Extrapolated to I = 0	KIN	S6
					For values of pK in D_2O/H_2O mixtures, see S6		
					Other measurements: B123, G64, K57, M80, N48		
123. Hyposulphurous acid, $H_2S_2O_4$							
0.35		2.45		25		C1	J3
124. Imidodiphosphoric acid, $H_2O_3P.NH.PO_3H_2$							
~1.5	2.66	7.32	10.22	25	Concentration constants; I = 0.1(NMe_4Br); $f_\pm$ assumed same as for HBr; pK_1 may be seriously in error because of experimental difficulties	E3bg	I13
~2	2.85	7.08	9.72	25	I = 0.2		
~2	2.81	7.05	9.77	25	I = 0.3		
~1.5	3.05	7.62	10.36	25	I = 1.0		
~1.8	2.60	7.16	9.79	37	I = 0.1; as above		I12
~1.8	2.68	6.99	9.52	37	I = 0.3		
~1.8	2.81	6.90	9.41	50	I = 0.1		
~1.8	2.83	6.88	9.32	50	I = 0.3		
125. Imidodisulphuryl fluoride $HN(SO_2F)_2$							
1.28				25		E3bg	R50

126. (Aquo) Indium(III) ion, In^{3+}						
4.43	3.9		25	Successive pK values for hydrolysis of In^{3+} to $InOH^{2+}$ and $In(OH)_2^+$; I = 3($NaClO_4$); using In-Hg electrode; above 0.001 M, indium forms $In[(OH)_2In]_n^{(3+n)+}$		B57
4.4	4.4		25	Successive pK values for hydrolysis of In^{3+}; I = 3($NaClO_4$); tracer amounts of In^{3+}	DISTRIB	R42
11.89	11.55	11.32	20±2	Successive pK_b values for hydrolysis of In^{3+} to $InOH^{2+}$, $In(OH)_2^+$ and $In(OH)_3$; I = 1	DISTRIB	H19
6.95			25	pK for hydrolysis of In^{3+} to form a mixed hydroxy-chloro complex; I = 3(NaCl); c = 0.001-0.04 M In^{3+}; a binuclear $In_2(OH)_2$ chloro complex is also formed	E3ag	B65
3.54			25	$I \to 0$; In/Hg electrode	E	Y4
3.54	4.28	5.16		Successive pK values for the hydrolysis of In^{3+} to $InOH^{2+}$, $In(OH)_2^+$ and $In(OH)_3$; I = 0	O8	B73
4.22			25		E	K105
4.22	7.16		25	In/Hg electrode		
				Other measurements: H48, H59, L11, M85, M86		
127. Iodic acid, HIO_3 (H_5IO_6)						
0.804			25	Obtained by three independent methods, taking ion-size parameter of 5A; value depends on ion-size assumed; emf method due to A.K. Covington and J.E. Prue, J. Chem. Soc., 1955, 3701	C1,E, and KIN	P42
0.785			25	Solubility of $AgIO_3$ in HNO_3 and KNO_3, extrapolated against $I^{1/2}$, I = 0.008 – 0.5	SOLY	L33
0.815			30			
0.84			35			
0.788			25	Solubility of $Ba(IO_3)_2$ in 1:1 electrolyte solutions, extrapolated against $I^{1/2}$; I = 0.0025 to 1	SOLY	N7

Name, Formula and pK values	T(^{o}C)	Remarks	Methods	Reference
0.773	25	I = 0.0026 to 0.01; extrapolated against $I^{1/2}$	O4	H14
0.773	25	Calculated from data of C.A. Kraus and H.C. Parker, J. Am. Chem. Soc., 44, 2429 (1922)	Cl, Rlc	F54
0.807	25	Calculated from data of C.A. Kraus and H.C. Parker, J. Am. Chem. Soc., 44, 2429 (1922), taking an ion size of 3A	Cl, Rlc	L22
0.58	0		FP	A4
0.72	18		Cl	
0.74	30	For I = 0	NMR	H86
1.01	25	pK values in formamide	SOLY	D6
0.92	30			
0.91	35			
		For values of pK in D_2O/H_2O mixtures, see S6 For H_0 acidity function of aqueous HIO_3, see D29, D30 For thermodynamics of ionization, see W36 Neutral aggregates of iodic acid are formed in aqueous solutions; see G53 Other measurements: D71, K29, K87, M6, O20, R47		
128. Iodine tetrafluoride hydroxy oxide, $HOIOF_4$				
5.0		pK in glacial acetic acid	O1	E29
129. (Aquo) Iridium(III) ion, Ir^{3+}				
4.78 5.63	5	Stepwise pK values for hydrolysis of Ir^{3+}; I = 1.05 ($NaClO_4$)	E3bg,h	G7
4.59 5.27	15			
4.37 5.20	25			
4.13 5.08	35			

130. (Aquo) Iron(II) ion, Fe^{2+}

6.93			20	p*K* for hydrolysis of Fe^{2+}; *I* = 0.5-2($NaClO_4$)	E3bg	B90
6.74			25			
6.49			35			
6.34			40			
7.15			20	p*K* for hydrolysis of Fe^{2+}; *I* = 1($NaClO_4$)	E3bg	B81
6.8			25			
8.3			25	p*K* for hydrolysis of Fe^{2+}	SOLY	L24
7.9			25	p*K* for hydrolysis of Fe^{2+}; *c* = 0.02-0.08 M $FeCl_2$; hydrolysis of "pure" salts	E3ag	G22
7.2				Concentration constants: *I* = 0.5(KCl)		
3.3			25	p*K* for hydrolysis of Fe^{2+}; from rate of H_2O_2 decomposition as function of pH in presence of Fe^{2+}; *I* = 1($NaClO_4$)	KIN	W19
				log *K* for $Fe^{2+} + 3OH^- \rightleftharpoons Fe(OH)_3^-$ is estimated from polarography to be 7.85 in 1.375 N NaOH		S40
7.34	13.33	16.95	18	−log *K* for $Fe^{2+} + nH_2O \rightleftharpoons Fe(OH)_n^{(2-n)-} + nH^+$; *I* = 0.1($NaClO_4$), using radionuclide	ION	I11
6.51	5.0		25	Successive p*K* values for the hydrolysis of Fe^{2+}	E,g	K64
8.50			12	*I* = 1($NaClO_4$)	E3bg	B88
8.03			15			
7.15			20			
6.8			25			
5.95			30			
9.58			25	p*K* for hydrolyis of Fe^{2+}; titration of $FeCl_2$	E3bg	M59
9.3			25		SOLY	S129
9.5			20	titration of $Fe(ClO_4)_2$; in 1M $NaClO_4$; calculating results by B.O.A. Hedstrom, Arkiv Kemi , 5, 457 (1953)		
8.30				p*K* for hydrolysis of Fe^{2+}; *I* = 2$(NH_4)_2SO_4$	O	E9
8.07				*I* = 2($NaClO_4$)		

Name, Formula and pK values		$T(^{o}C)$	Remarks	Methods	Reference
25.70			at pH 12-14; $-\log \underline{K}$ for $FeOH^{+} + H_2O \rightleftharpoons FeO_2H^{-} + 2H^{+}$ Other measurements: H51, J10, L40c		
131. (Aquo) Iron(III) ion, Fe^{3+}					
2.71		15	$p\underline{K}$ for hydrolysis of Fe^{3+}; concentration constant; $\underline{I} = 0.01$	O5	T33
2.46		25			
2.29		35			
2.30		20	$p\underline{K}$ for hydrolysis of Fe^{3+}; $\underline{I} = 0.025$ to 0.15	O6	R21
2.34		25	($NaClO_4$, $HClO_4$); extrapolated to $\underline{I} = 0$		
2.38		18	$p\underline{K}$ for hydrolysis of Fe^{3+}; $\underline{I} = 0.01$ to 0.03; extrapolated	O6	M73
2.19		25	to $\underline{I} = 0$		
2.02		32			
2.96		18	$\underline{I} = 1(NaClO_4)$; constants are also given for $2FeOH^{2+} \rightleftharpoons$		
2.79		25	$Fe_2(OH)_2^{4+}$		
2.61		32			
2.17		25	$p\underline{K}$ for hydrolysis of Fe^{3+}; $\underline{I} = 0.015$ to 3.0; extrapolated to $\underline{I} = 0$ using Debye-Hückel equation; constants are also given for $2FeOH^{2+} \rightleftharpoons Fe_2(OH)_2^{4+}$	O6	M74
2.19		25	For $\underline{I} = 0$	O6	S75
2.63		20-22	$\underline{I} = 0.1(KNO_3)$	O5	P57
2.80		25	$\underline{I} = 0.5(NaClO_4)$	O6	W27
2.92		21	$\underline{I} = 0.55$; in D_2O	O6	H95
2.74	3.31	20	Successive $p\underline{K}$ values for hydrolysis of Fe^{3+}; $\underline{I} = 1(NaClO_4)$; $-\log \underline{K} = 2.85$ for $2Fe^{3+} + 2H_2O \rightleftharpoons Fe_2(OH)_2^{4+} + 2H^{+}$	REDOX	P33
2.83	4.59	25	Successive $p\underline{K}$ values for hydrolysis of Fe^{3+}	E	I20
2.83		25		C	
3.00		25	$p\underline{K}$ for $Fe^{3+} + H_2O \rightleftharpoons FeOH^{2+} + H^{+}$; $\underline{I} = 3(NaClO_4)$	REDOX	Z8
2.73		35			
2.52		45			

2.36			25	$-\log K$ for $2Fe^{3+} + 2H_2O \rightleftharpoons Fe_2(OH)_2^{4+} + 2H^+$		
2.17			35			
2.03			45			
2.93			25	$-\log K$ for $Fe^{3+} + H_2O \rightleftharpoons FeOH^{2+} + H^+$; $I = 0.05$	O5	S13
2.44			55			
2.06			80			
2.18			25	$-\log K$ for $Fe^{3+} + H_2O \rightleftharpoons FeOH^{2+} + H^+$; $I = 0.1$-2.0,	O	Z6
1.62			50	extrapolated to $I = 0$; $c = 10^{-4}$ M $Fe(ClO_4)_2$		
1.02			80			
1.00			80	$-\log K$ for $Fe^{3+} + H_2O \rightleftharpoons FeOH^{2+} + H^+$; $I = 0.1$-2.0,		Z7
0.51			110	extrapolated to $I = 0$; $c = 5.10^{-4}$ $Fe(ClO_4)_3$		
0.05			140			
-0.30			170			
-0.66			200			
3.1			25	$-\log K$ for $Fe^{3+} + H_2O \rightleftharpoons FeOH^{2+} + H^+$; $I = 3(NaClO_4)$	E3bg	C35
2.8			25	$-\log K$ for $2Fe^{3+} + 2H_2O \rightleftharpoons Fe_2(OH)_2^{4+} + 2H^+$; $I = 3(NaClO_4)$		
46.1			25	$-\log K$ for $12Fe^{3+} + 34H_2O \rightleftharpoons Fe_{12}(OH)_{34}^{2+} + 34H^+$; $I = 3(NaClO_4)$		
3.15			25	$-\log K$ for $Fe^{3+} + D_2O \rightleftharpoons FeOD^{2+} + D^+$	O2	K44
3.45			25	$-\log K$ for $Fe^{3+} + H_2O \rightleftharpoons FeOH^{2+} + H^+$		
11.66			18	$\log K$ for $Fe^{3+} + OH^- \rightleftharpoons FeOH^{2+}$	ION	K48
15.93				$\log K$ for $2Fe^{3+} + OH^- \rightleftharpoons Fe_2OH^{5+}$		
22.07				$\log K$ for $Fe^{3+} + 2OH^- \rightleftharpoons Fe(OH)_2^+$		
26.73				$\log K$ for $2Fe^{3+} + 2OH^- \rightleftharpoons Fe_2(OH)_2^{4+}$		
30.85				$\log K$ for $3Fe^{3+} + 2OH^- \rightleftharpoons Fe_3(OH)_2^{7+}$		
30.70				$\log K$ for $Fe^{3+} + 3OH^- \rightleftharpoons Fe(OH)_3$		
34.48				$\log K$ for $2Fe^{3+} + 3OH^- \rightleftharpoons Fe_2(OH)_3^{3+}$		
12.02	23.72	35.23		$\log K$ for $Fe^{3+} + nOH^- \rightleftharpoons Fe(OH)_n^{(3-n)+}$	O	K72
3.05	3.26		25	Successive pK values for hydrolysis of Fe^{3+}; $I = 3(NaClO_4)$;	REDOX	H52

Name, Formula and pK values	T(°C)	Remarks	Methods	Reference
		also $-\log \underline{K} = 2.91$ for $2Fe^{3+} + 2H_2O \rightleftharpoons Fe_2(OH)_2^{4+} + 2H^+$		
		Values of $-\log \underline{K}$ for $2Fe^{3+} + 2H_2O \rightleftharpoons Fe_2(OH)_2^{4+} + 2H^+$, from 15-41°, are estimated from magnetic measurements		M109
		Other measurements: A44, B12, B77, B78, B107, B125, B128, C44, D12, I25, L6, L40b, O18, S81, S125, V11, Y5		
132. Isohypophosphoric acid, $H_4P_2O_6$				
4.5 8.5		$p\underline{K}_2$, $p\underline{K}_3$; $c = 0.02M$	E3bg	B85
1.67 6.26	25	$p\underline{K}_2$, $p\underline{K}_3$; $\underline{I} = 0.1$ to 1.0 (Et_4NCl); extrapolated to $\underline{I} = 0$; $p\underline{K}_1$ estimated as 0.6	E3bg	C11
133. (Aquo) Lanthanum(III) ion, La^{3+}				
~10	25	$p\underline{K}$ for hydrolysis of La^{3+}; from hydrolysis of "pure" salt; $c = 0.001-0.01$ M $La_2(SO)_3$	E3ag	M87
9.06	25	$p\underline{K}_a$ for hydrolysis of La^{3+}; titration of 0.004-0.009 M $La(ClO_4)_3$ with 0.02 M $Ba(OH)_2$; $\underline{I} = 0.3$ ($NaClO_4$)	E3b	F49
8.98	25	ditto, using 0.02 M NaOH		
10.1	25	$p\underline{K}$ for hydrolysis of La^{3+}; $\underline{I} = 3(LiClO_4)$; $c = 0.1-1.0$ M $La(ClO_4)_3$; also $-\log \underline{K} = 9.95$ for $2La^{3+} + H_2O \rightleftharpoons La_2OH^{5+} + H^+$; other species include $La_5(OH)_9^{6+}$ and $La_6(OH)_{10}^{8+}$	E3bg	B60
5.6	20	$p\underline{K}_b$; $c = 0.01$ M $LaCl_3$	E3bg	W23
3.3	25	$p\underline{K}_b$; estimated from solubility measurements of I.M. Kolthoff and R. Elmquist, J. Am. Chem. Soc., 53, 1217 (1931)		D18
~5	18	$p\underline{K}_b$	DISTRIB	V12
9.06	25	$p\underline{K}$ for hydrolysis of La^{3+}; $\underline{I} = 0$	E3bg	U2
10.04	25	$p\underline{K}$ for hydrolysis of La^{3+}; $\underline{I} = 3(LiClO_4)$	E3bg	A26
10.35	25	$p\underline{K}$ for hydrolysis of La^{3+} in D_2O solution; $\underline{I} = 3(LiClO_4)$	E3bg	A27
		For $p\underline{K}$ values of lanthamides, see G74		

134. (Aquo) Lead(II) ion, Pb^{2+}

7.86			25	p$\underline{K}$ for hydrolysis of Pb^{2+}; $\underline{I}$ = 0.1(KNO_2); [Pb^{2+}] = 0.1-2.0.$10^{-3}$$\underline{M}$	E3bg	S131
				Also 23.91 = –log $\underline{K}$ for $3Pb^{2+} + 4H_2O \rightleftharpoons Pb_3(OH)_4^{2+} + 4H^+$		
				36.75 = –log $\underline{K}$ for $3Pb^{2+} + 5H_2O \rightleftharpoons Pb_3(OH)_5^{+} + 5H^+$		
				20.40 = –log $\underline{K}$ for $4Pb^{2+} + 4H_2O \rightleftharpoons Pb_4(OH)_4^{4+} + 4H^+$		
				43.38 = –log $\underline{K}$ for $6Pb^{2+} + 8H_2O \rightleftharpoons Pb_6(OH)_8^{4+} + 8H^+$		
7.9				p$\underline{K}$ for hydrolysis of Pb^{2+}		G25
5.7				–log $\underline{K}$ for $PbOH^+ + H_2O \rightleftharpoons Pb(OH)_2 + H^+$		
14.3				log $\underline{K}$ for $HPbO_2^- + H^+ \rightleftharpoons Pb(OH)_2$		
7.78			18	p$\underline{K}$ for hydrolysis of Pb^{2+}; for $\underline{I}$ = 0; c = 0.005-0.4 M $(PbNO_3)_2$; also log $\underline{K}$ = –7.30 for $2Pb^{2+} + H_2O \rightleftharpoons Pb_2OH^{3+} + H^+$, and log $\underline{K}$ = –20.93 for $4Pb^{2+} + 4H_2O \rightleftharpoons Pb_4(OH)_4^{4+} + 4H^+$	E3bg	P28
8.66			20	p$\underline{K}$ for hydrolysis of Pb^{2+}; $Pb_4(OH)_4^{4+}$ is also formed	E3bg	F5
7.93			25	p$\underline{K}$ for hydrolysis of Pb^{2+}; $\underline{I}$ = 2($NaClO_4$); also log $\underline{K}$ = – 19.35 for $4Pb^{2+} + 4H_2O \rightleftharpoons Pb_4(OH)_4^{4+} + 4H^+$		H96
8.84			25	p$\underline{K}$ for hydrolysis of Pb^{2+}; $\underline{I}$ = 2($NaNO_3$); also log $\underline{K}$ –7.11 for $2Pb^{2+} + H_2O \rightleftharpoons Pb_2OH^{3+} + H^+$; log $\underline{K}$ = –21.72 for $4Pb^{2+} + 4H_2O \rightleftharpoons Pb_4(OH)_4^{4+} + 4H^+$	E3ag	H97
7.1	10.1	11.5	25	p$\underline{K}$ values for stepwise hydrolysis of Pb^{2+} to $PbOH^+$,$(PbOH)_2$ and $(Pb(OH)_3^-$; $\underline{I}$ = 1(KNO_3)	POLAROG	G51
7.8	9.4	10.8	25	p$\underline{K}$ values for stepwise hydrolysis of Pb^{2+}; $\underline{I}$ = 0.3($NaClO_4$); Pb-Hg electrode		C6
7.9	9.6	11.5	25	$\underline{I}$ = 3($NaClO_4$)		
7.8			25	p$\underline{K}_b$ for $PbOH^+ \rightleftharpoons Pb^{2+} + OH^-$	SOLY	G13
5.99			100	p$\underline{K}$ for hydrolysis of Pb^{2+}	KIN	K108
				At high lead concentrations, Pb^{2+} also hydrolyses to Pb_2OH^{3+} and $Pb_4(OH)_4^{3+}$ (constants are given)		O15,P5
				At 25° and $\underline{I}$ = 2($NaClO_4$), log $\underline{K}$ = 12.62 for $Pb^{2+} + 3OH^- \rightleftharpoons Pb(OH)_3^-$	POLAROG	O19
				At 25° and $\underline{I}$ = 0, log $\underline{K}$ = 13.90 for $Pb^{2+} + 3OH^- \rightleftharpoons Pb(OH)_3^-$	POLAROG	N49

Name, Formula and pK value	T(°C)	Remarks	Methods	Reference
		At 25° and $\underline{I}$ = 0, log $\underline{K}$ = 13.95 for $Pb^{2+} + 3OH^- \rightleftharpoons Pb(OH)_3^-$	POLAROG	V18
		At 20°, log $\underline{K}$ = 12.15 for $Pb^{2+} + 3OH^- \rightleftharpoons Pb(OH)_3^-$	POLAROG	H62
		Other measurements: B71, C58, G77, G79, K47, S39, T20, W34		
135. (Aquo) Lead(IV) ion, Pb^{4+}				
1.8		–log $\underline{K}$ for $Pb^{4+} + H_2O \rightleftharpoons Pb(OH)^{3+} + H^+$		G25
3.2		–log $\underline{K}$ for $Pb^{4+} + 2H_2O \rightleftharpoons Pb(OH)^{2+} + 2H^+$		
5.2		–log $\underline{K}$ for $Pb^{4+} + 3H_2O \rightleftharpoons Pb(OH)_3^+ + 3H^+$		
6.7		–log $\underline{K}$ for $Pb^{4+} + 4H_2O \rightleftharpoons Pb(OH)_4 + 4H^+$		
136. (Aquo) Lithium ion, Li^+				
0.26	5	$p\underline{K}_b$; $\underline{I}$ = 0.02 to 0.1; $f\pm$ calculated using Davies' equation; e.m.f. data from H.S. Harned and H.R. Copson, J. Am. Chem. Soc., 55, 2206 (1933), and H.S. Harned and J.G. Donelson, J. Am. Chem. Soc., 59, 1280 (1937)	Elch	G40
0.20	15			
0.18	25			
0.20	35			
0.19	45			
–0.08	25	$p\underline{K}_b$; $f\pm$ from Davies' equation	C2,R1b	D5
–0.53	25	$p\underline{K}_b$	C2,R1d	S102
0.32	25	$p\underline{K}_b$	C2,R1e	
–0.18	25	$p\underline{K}_b$; concentration constant; $\underline{I}$ = 3($NaClO_4$); taking $f OH^- = f Cl^-$	E2ah	O4
0.36		$p\underline{K}_b$; concentration constant; from salt effect on indicator; $\underline{I}$ = 1(LiCl)	O3	K54
0.13		$\underline{I}$ = 0.2(LiCl)		
0.89	49	$p\underline{K}_b$; for $\underline{I}$ = 0	C2	W39
1.13	93			

1.51	138			
1.42	182			
1.59	227			
1.76	271			
		For alkalinity function of LiOH solutions, see L16, M89, S112a		
		For H_- acidity function of concentrated LiOH solutions see Y1		

137. (Aquo) Lutecium(III) ion, Lu^{3+}

6.6	20	pK_b for $LuOH^{2+} \rightleftharpoons Lu^{3+} + OH^-$; c = 0.01 M $LuCl_3$	E3bg	W23
7.90	25	pK_a for hydrolysis of Lu^{3+}; titration of 0.004-0.009 M $Lu(ClO_4)_3$ with 0.02 M $Ba(OH)_2$; I = 0.3($NaClO_4$)	E3b	F49
7.98	25	ditto, using 0.02 M NaOH		
7.66	25	I = 0	E3bg	U2

138. (Aquo) Magnesium ion, Mg^{2+}

2.58	25	pK_b; for I = 0; c = 0.03 M $MgCl_2$	E3bg	S113
2.1	18	pK_b	SOLY	G42
2.60	25	pK_b; I = 0	E3bg	H91
2.4	18	pK_b; concentration constant; c = 0.1-0.5 N $MgCl_2$; salt effect on indicator	O3	K55
12.2	25	pK for hydrolysis of Mg^{2+}; I = 3(NaCl, $MgCl_2$)	E3bg,h	L30
12.8	30	pK for hydrolysis of Mg^{2+}, I = 0.1(KCl); c = 0.01 M	E3bg	C14
9.76	100	pK for hydrolysis of Mg^{2+}; taking pK_w = 12.38; c = 0.06 M $MgCl_2$; rate of inversion of sucrose	KIN	K108
10.5	100	$-\log K$ for $Mg^{2+} + H_2O \rightleftharpoons MgOH^+ + H^+$; I = 3($NaClO_4$);	E	G9
9.9	140	c = 0.4-1.4M $Mg(ClO_4)_2$		
10.0	100	$-\log K$ for $2Mg^{2+} + H_2O \rightleftharpoons Mg_2OH^{3+} + H^+$; I = 3($NaClO_4$)		
9.0	140			

Name, Formula and pK value	T(^{o}C)	Remarks	Methods	Reference
31.0	100	$-\log \underline{K}$ for $4Mg^{2+} + 4H_2O \rightleftharpoons Mg_4(OH)_4^{4+} + 4H^+$; $\underline{I} = 3(NaClO_4)$		
27.0	140			
12.00	25	$-\log \underline{K}$ for $Mg^{2+} + H_2O \rightleftharpoons MgOH^+ + H^+$; $\underline{I} = 3(NaClO_4)$; c = 0.6-1.4M $Mg(ClO_4)_2$	Eg	B144
11.70	40			
11.25	60			
10.5	100			
8.9	140			
12.30	25	$-\log \underline{K}$ for $2Mg^{2+} + H_2O \rightleftharpoons Mg_2(OH)^{3+} + H^+$; $\underline{I} = 3(NaClO_4)$; c = 0.6–1.4M $Mg(ClO_4)_2$	Eg	B144
11.75	40			
11.10	60			
10.0	100			
9.0	140			
38.30	25	$-\log \underline{K}$ for $4Mg^{2+} + 4H_2O \rightleftharpoons Mg_4(OH)_4^{4+} + 4H^+$; $\underline{I} = 3(NaClO_4)$; c = 0.6–1.4M $Mg(ClO_4)_2$		
36.85	40			
34.00	60			
31.00	100			
27.00	140			
		For other measurements giving polymeric species, see E12		
139. (Aquo) Manganese(II) ion, Mn^{2+}				
10.93	15	p$\underline{K}$ for hydrolysis of Mn^{2+}; $\underline{I}$ = 0.002 to 0.04; extrapolated to $\underline{I}$ = 0 by fitting extended Debye-Hückel equation	E3bg	P35
10.76	20			
10.59	25			
10.38	30			
10.19	36			
10.10	42			

10.6		30	p*K* for hydrolysis of Mn^{2+}; *I* = 0.1(KCl)	E3bg	C14
9.54		100	p*K* for hydrolysis of Mn^{2+}	KIN	K108
10.5		25	p*K* for hydrolysis of Mn^{2+}; *I* = 1($NaSO_4$); also $-\log K = 9.9$ for $2Mn^{2+} + H_2O \rightleftharpoons Mn_2(OH)^{3+} + H^+$; $-\log K = 25.4$ for $3Mn^{2+} + 2H_2O \rightleftharpoons Mn_3(OH)_2^{4+} + 2H^+$		F34
140. (Aquo) Manganese(IV) ion, Mn^{3+}					
0.20		12.4	p*K* for hydrolysis of Mn^{3+}; *I* = 3($NaClO_4$)	KIN	R40
0.4		25	$-\log K$ for $Mn^{3+} + H_2O \rightleftharpoons MnOH^{2+} + H^+$	REDOX	B67
0.1			$-\log K$ for $Mn^{3+} + 2H_2O \rightleftharpoons Mn(OH)_2^+ + 2H^+$		
0.06		25	p*K* for hydrolysis of Mn^{3+}; *I* = 4($Mn(ClO_4)_2$,$HClO_4$)	O6	W18
~−0.2		23	*I* = 5.3 to 6.1 ($Mn(ClO_4)_2$,$HClO_4$)	O6	F1
~−0.7		23	*I* = 6($HClO_4$, $NaClO_4$)	O6	D52
141. Manganic acid, H_2MnO_4					
10.15		35	pK_2; *I* ≈ 0.1	KIN	L48
12.5		20	Estimate		H50
142. (Aquo) Mercury(I) ion, Hg_2^+					
4.88			$-\log K$ for $Hg_2^{2+} + H_2O \rightleftharpoons Hg_2(OH)^+ + H^+$; *I* = 3($NaClO_4$)	Hg,g	H64
2.68			$-\log K$ for $2Hg_2^{2+} + H_2O \rightleftharpoons (Hg)_2OH^{3+} + H^+$		
5.0		25	p*K* for hydrolysis of Hg_2^{2+}; *I* = 0.5($NaClO_4$); measurements included Hg electrode; allowed for equilibrium, $Hg + Hg^{2+} \rightleftharpoons Hg_2^{2+}$; earlier reported values are too low because of the hydrolysis of Hg^{2+} also present	E3bg	F36
4.6			*c* = 0.006 M, as perchlorate	E3ag	N33
143. (Aquo) Mercury(II) ion, Hg^{2+}					
3.49	2.47	25	Successive p*K* values for the hydrolysis of Hg^{2+}; *I* = 3($Ca(ClO_4)_2$,$Mg(ClO_4)_2$)	E3bg	A16
3.55	2.66	25	*I* = 3($NaClO_4$)		

Name, Formula and pK values		T(^{o}C)	Remarks	Methods	Reference
			Constants are also given for Hg_2OH^{3+}, $Hg_2(OH)_2^{2+}$ and $Hg_4(OH)_3^{5+}$		
3.70	2.65	25	$\underline{I}$ = 0.5($NaClO_4$)	E3bg	H65
3.23	2.93	25	$\underline{I}$ = 3($NaClO_4$)	SOLY	D75
2.49	2.85	25		SOLY	G12
6.72		13	–log $\underline{K}$ for $Hg^{2+} \rightleftharpoons Hg(OH)_2 + 2H^+$; $\underline{I}$ = 0.1($NaNO_3$)	E3bg	A32
6.52		20			
6.26		30			
6.00		40			
6.22		20	$\underline{I}$ = 0, by extrapolation against $\underline{I}^{1/2}$		
			Ratio of successive constants for hydrolysis of Hg^{2+} is about 0.04		
6.26		25	–log $\underline{K}$ for $Hg^{2+} \rightleftharpoons Hg(OH)_2 + 2H^+$	SOLY	G10
14.85		25	p$\underline{K}$ for $Hg(OH)_2 + H_2O \rightleftharpoons Hg(OH)_3^- + H^+$		
14.77		25	p$\underline{K}$ for $Hg(OH)_2 + H_2O \rightleftharpoons Hg(OH)_3^- + H^+$	SOLY	F56
21.4		30	log $\underline{K}$ for $Hg^{2+} + 2OH^- \rightleftharpoons Hg(OH)_2$; $\underline{I}$ = 2($NaNO_3$)	POLAROG	N34
			Other measurements: B81, D2, G58, G77, K49		
144. (Aquo) Molybdenum (III) ion, Mo^{3+}					
12.3		20	log $\underline{K}$ for $Mo^{3+} + OH^- \rightleftharpoons MoOH^{2+}$ $\underline{I}$ = 1(NaOH,HCl,NaCl)	O8	M82
23.4			log $\underline{K}$ for $Mo^{3+} + 2OH^- \rightleftharpoons Mo(OH)_2^+$		
34.7			log $\underline{K}$ for $Mo^{3+} + 3OH^- \rightleftharpoons Mo(OH)_3$		
145. Molybdic acid, H_2MoO_4			See also Heptamolybdic acid, Tetramolybdic acid		
			Acidification of molybdate solutions gives polymeric species, of which $Mo_7O_{24}^{6-}$ is believed to be the simplest to be formed in appreciable amounts. See A60, G45		
	4.08	25	Concentration constant; $\underline{I}$ = 3($NaClO_4$); formation of $Mo_7O_{24}^{6-}$ is important even down to $c = 6 \times 10^{-4}$M	Eleg,h	S17

~3.6	3.89	25	I = 3($NaClO_4$); more refined values from data given in S17	Elcg,h	S18
4.00	4.21	20	I = 0.0023; $c = 10^{-4}$ M molybdate	O5	R33
3.52	4.84	22	I = 0.1; from electromigration in $NaNO_3$ solutions		C29
3.57	4.75	22	I = 0.1; from electromigration in $NaClO_4$ solutions		
1.15	3.75	21	I = 0.465	O5	Y14
0.9		25	pK for proton addition: I = 0.5($NaClO_4$, $HClO_4$) $c = 10^{-4}$ M molybdate	O6	R34
0.3		22	pK for proton addition, from electromigration in $NaNO_3$ solutions		C29
0.8		22	pK for proton addition, $NaClO_4$ solutions		
			Other measurements: I26, N44, S96		
3.57		25	log K for $H^+ + MoO_4^{2-} \rightleftharpoons HMoO_4^-$; I = 1(NaCl); c = 0.0005–0.10M sodium molybdate	E3bg	C63
7.28			log K for $2H^+ + MoO_4^{2-} \rightleftharpoons H_2MoO_4$		
52.79			log $\beta_{8,7}$ for $8H^+ + 7MoO_4^{2-} \rightleftharpoons H_8(MoO_4)_7^{6-}$		
57.48			log $\beta_{7,7}$ for $9H^+ + 7MoO_4^{2-} \rightleftharpoons H_9(MoO_4)_7^{5-}$		
61.04			log $\beta_{10,7}$ for $10H^+ + 7MoO_4^{2-} \rightleftharpoons H_{10}(MoO_4)_7^{4-}$		
63.52			log $\beta_{11,7}$ for $11H^+ + 7MoO_4^{2-} \rightleftharpoons H_{11}(MoO_4)_7^{3-}$		
2.47	3.80	18	Stepwise pK values for ionization of H_2MoO_4; I = 0.3–1.0	O	N2
1.0		18	log K for $HMoO_3^+ + H^+ \rightleftharpoons MoO_2^{2+} + H_2O$		
2.0		18	log K for $H_2MoO_4 + H^+ \rightleftharpoons HMoO_3^+ + H_2O$		
			In alkaline solution, Mo(VI) exists as MoO_4^{2-} but in acid or neutral solutions polynuclear species are the main products		S19
146. Monobromamine, NH_2Br					
6.39			I = 0	O5	J7

Name, Formula and pK values	T(°C)	Remarks	Methods	Reference
147. Monochloramine, NH_2Cl				
15	25	pK_b; estimate based on pK-lowering by chlorine substitution in dialkylamines		W15
148. (Aquo) Neodymium(III) ion, Nd^{3+}				
8.5	25	pK for hydrolysis of Nd^{3+} to $NdOH^{2+}$; $I = 3(NaClO_4)$	E3b, quin	T13
~9	25	pK for hydrolysis of Nd^{3+}; hydrolysis of "pure" satls; $c = 0.001 - 0.01$ M $Nd_2(SO_4)_3$	E3ag	M87
8.43	25	pK_a for hydrolysis of Nd^{3+}; titration of 0.004-0.009 M $Nd(ClO_4)_3$ with 0.02 M $Ba(OH)_2$; $I = 0.3(NaClO_4)$	E3b	F49
9.4	25	pK for hydrolysis of Nd^{3+} to $NdOH^{2+}$; $I = 3(NaClO_4)$; $c = 0.2-0.8$ Nd^{3+}	E	B153
13.93		$-\log K$ for $2Nd^{3+} + 2H_2O \rightleftharpoons Nd_2(OH)_2^{4+} + 2H^+$		
11.86		$\log K_{2,1}$, in H_2O		
12.28		$\log K_{2,1}$ in D_2O		
15.90		$-\log {}^*\beta_{2,1}$ in H_2O		
16.54		$-\log {}^*\beta_{2,1}$ in D_2O		
149. (Aquo) Neptunium(III) ion, Np^{3+}				
7.43	25	pK for hydrolysis of Np^{3+} to $NpOH^{2+}$; $I = 0.3$; $c = 0.044M$ Np^{3+}	E	M51
150. (Aquo) Neptunium(IV) ion, Np^{4+}				
2.30	25	pK for hydrolysis of Np^{4+}; $I = 2(NaClO_4)$	O5	S120
2.49	25	$I = 2(NaClO_4)$; D_2O solution. At low acidities, polymerization becomes important. Other measurements: D70, M100		

151. (Aquo) Neptunium(V) ion, Np^{5+}				
8.9		p*K* for $NpO_2^+ + H_2O \rightleftharpoons NpO_2OH + H^+$		K84
8.91		in HNO_3	E	S61
8.88		in $HClO_4$	E	
8.89			O	
		Other measurements: M100		
152. (Aquo) Neptunium(VI) ion,				
		See M100		
153. (Aquo) Nickel(II) ion, Ni^{2+}				
10.22	15	p*K* for hydrolysis of Ni^{2+}; *I* = 0.0016 to 0.043; extrapolated to *I* = 0 by fitting to extended Debye-Hückel equation	E3bg	P37
10.05	20			
9.86	25			
9.75	30			
9.58	36			
9.43	42			
8.94	25	p*K* for hydrolysis of Ni^{2+}; *I* = 0.007 to 0.55, extrapolated to *I* = 0	E3bg	C67
9.23-9.49	25	p*K* for hydrolysis of Ni^{2+}, as nitrate, sulphate, chloride perchlorate	E3bg	K98
9.4	30	*I* = 0.1(KCl)	E3bg	C14
8.60	100		KIN	K108
10.01	28	p*K* for hydrolysis of Ni^{2+}; *I* = 1($NaClO_4$)	E3bg	S63
9.76	25	p*K* for hydrolysis of Ni^{2+}; *I* = 1($NaClO_4$)	E3bg	B91
9.65	30			
9.52	40			
9.3	50			
28.42	25	$-\log K$ for $4Ni^{2+} + 4H_2O \rightleftharpoons Ni_4(OH)_4^{4+}$; *I* = 3(NaCl)	E3bg	B151
9.3		$-\log K$ for $2Ni^{2+} + H_2O \rightleftharpoons Ni_2OH^{3+} + H^+$		

Name, Formula and pK values	T(°C)	Remarks	Methods	Reference
21.58	25	−log K for $3Ni^{2+} + 3H_2O \rightleftharpoons Ni_3(OH)_3^{3+} + 3H^+$; I = 3($NaNO_3$)	E3bg	B149
9.6		−log K for $2Ni^{2+} + H_2O \rightleftharpoons Ni_2OH^{3+} + H^+$		
27.32	25	−log K, $4Ni^{2+} + 4H_2O \rightleftharpoons Ni_4(OH)_4^{4+} + H^+$; I = 3($LiClO_4$)	E1a	K18
13.3		−log K for $Ni^{2+} + 2H_2O \rightleftharpoons Ni(OH)_2 + 2H^+$		
28.18	25	−log K for $4Ni^{2+} + 4H_2O \rightleftharpoons Ni_4(OH)_4^{4+} + 4H^+$; I = 3(NaBr)	E,g	B150
9.5		−log K for $2Ni^{2+} + H_2O \rightleftharpoons Ni_2OH^{3+} + H^+$		
28.5	25	−log K for $4Ni^{4+} + 4H_2O \rightleftharpoons Ni_4(OH)_4^{4+} + 4H^+$; I = 3(NaCl); c = 0.015-1.00 M in Ni^{2+}	E3bg	O7
10.24	25	pK for hydrolysis of Ni^{2+}; I = 3($LiClO_4$); c = 0.004-0.40 M $NO(ClO_4)_2$	E,h	K22
15.61		−log K for $2Ni^{2+} + 2H_2O \rightleftharpoons Ni_2(OH)_2^{2+} + 2H^+$		
9.51		−log K for $2Ni^{2+} + H_2O \rightleftharpoons Ni_2OH^{3+} + H^+$		
38.78		−log K for $2Ni^{2+} + 6H_2O \rightleftharpoons Ni_2(OH)_6^{2-} + 6H^+$; c = 10-40 in M Ni^{2+}		B55
		For hydrolysis constants in dioxane/water mixtures, see K18		
		At 25°, I = 3($NaClO_4$), and c = 0.1-0.8 M Ni^{2+}, the main hydrolysed species is $Ni_2(OH)_4^{4+}$		B154
		Other measurements: A6, D36, G21, M76, S44		
153a Niobic acid, $H_8Nb_6O_{19}$				
10.88 13.8	25	pK_7, pK_8; I = 3(KCl); c = 0.05-0.14 M in niobate	E3bh	N28
7.4	18-20	acidic pK		B1
14.6		basic pK		
3.22		−log K for $NbO_2^+ + H_2O \rightleftharpoons NbO_2OH + H^+$; I = 0.1($H^+$,$Li^+$)$ClO_4$	DISTRIB	G76
154. Nitramide, No_2NH_2				
6.48	25	Based on pK = 7.17 for o-nitrophenol	O2	T19

6.59	15	For I = 0: c = 0.002-0.03 M	C1	B126
6.5	~20		O5	

155. Nitric acid, HNO_3

−1.27	25	Molal scale; from vapour pressure and activity coefficient data for 2-14 M HNO_3	VAP	D27
−1.19	25	Molar scale		
−1.65	0		RAMAN	K88
−1.37	25			
−1.24	50			
−1.68	0		NMR	H88
−1.44	25			
−1.18	70			
−1.32	26±2		RAMAN	R16
−1.53	25	Molal scale; from vapour pressure and isopiestic	VAP	H2
−1.31	50	2-28 mol. $kg.^{-1}$		
−1.16	75			
−1.54	0	HNO_3 in H_2O	NMR	R17
−1.68		HNO_3 in D_2O		
−1.51		DNO_3 in D_2O		
−1.30	25	HNO_3 in H_2O		
−1.40		HNO_3 in D_2O		
−1.26		DNO_3 in D_2O		
−1.00	65	HNO_3 in H_2O		
−1.00		HNO_3 in D_2O		
−0.85		DNO_3 in D_2O		
−1.64	27±1	pK; molal scale	NMR	R9
−3.65	25		VAP	M30
0.55	100	molal scale; solubility of $CaSO_4$	SOL	M36
0.26	125			

Name, Formula and pK values	T(^{o}C)	Remarks	Methods	Reference
0.08	150			
−0.51	200			
−1.11	250			
−1.97	300			
−3.58	350			
−15.2	rt	p*K* of R-OH + $H^+ \rightleftharpoons R^+ + H_2O$, where R-OH = HNO_3; basicity of HNO_3 in 80-90% H_2SO_4	RAMAN	M42
−1.0 to −1.3	25	From vapour pressure and activity coefficient data	VAP	K7
−1.60		From dielectric constant; extrapolating from non-aqueous solutions		W44
−1.44	25	Assuming species formed is $HNO_3.3H_2O$	RAMAN and NMR	H77
−2.09	25	Calculated p*K* for a 1:1 HNO_3:H_2O species		
−3.78	25	Calculated p*K* for unhydrated nitric acid		
		Other measurements: B50, E29, H87, K42, K77, M11, N45, P38, R15, W43		
		Ref. H56 gives an equation fitting literature values of p*K* from 0° to 300°		
		For Hammett acidity function of HNO_3, see B16, D31, D33 (in presence of $LiNO_3$ and $NaNO_3$), L17 (in presence of $NaClO_4$), and P20		
		For H_R (Jo) acidity function of HNO_3, see D40		
		For $H_R{}^*$ acidity function of HNO_3, see Y9		
		For H_O acidity function, see D30, V16.		
156. Nitrous acid, HNO_2				
3.222	15	For *I* = 0, extrapolated from results in 0.01–4.0 *M* $NaNO_3$	E3bg	T30
3.162	20			
3.130	25			
3.078	35			

3.230		15	For $\underline{I} = 0$; extrapolated from results in 0.01 M $NaNO_2$, 0.03 M $NaNO_3$, and 0.21 M $NaClO_4$	E3bg	L62
3.203		20			
3.113		35			
3.49		0	$\underline{I} = 0.001$	O1	K38
3.34		12.5			
3.22		30			
3.46		0	$\underline{I} = 0.002$; using flowing solutions	O1	S30
3.29		25			
3.15		50			
3.148		25	$\underline{I} = 0.04$ to 2.0 ($NaClO_4$); extrapolated to $\underline{I} = 0$	E3bg	L63
3.15		20	"Practical" constant; $\underline{I} = 0.012$	E3bg	B117
3.26		25	For $\underline{I} = 0$; from $\underline{I} = 0.17$, assuming $\gamma_{+} = 0.773$	KIN	L35
3.29		25	"Practical" constant; $\underline{I} = 0.07(NaClO_4)$	O5	V7
2.80		25	$\underline{I} = 1(NaClO_4)$		
3.03		15	$\underline{I} = 1(NaNO_2, HClO_4, NaClO_4)$; concentration constant	O	H76, H76a
2.99		25			
2.94		35			
−8.1		25	p$\underline{K}$ for nitrosonium ion (NO^+) formation from HNO_2, using H_R function for $HClO_4$ solutions	O6	D41
−7.86		20	p$\underline{K}$ for nitrosonium ion (NO^+) formation from HNO_2, using C_o function for H_2SO_4 solutions	O6	S56
			Other measurements A47, B26, B84, D51, L34, R14, S42		

157. Octaminotetraphosphazene, $N_4P_4(NH_2)_8$

5.22	7.55	25	$\underline{I} = 0$	E	F11
5.15	7.50	25	$\underline{I} = 0$	E	F12

158. Orthophosphoric acid, see Phosphoric acid

159. Osmic acid, $H_4O_5O_6$

Name, Formula and pK values				T(°C)	Remarks	Methods	Reference
12.0	14.5			25	Apparent p*K* values; *I* = 1; the neutral species is mainly osmium tetroxide	O7	S12
12.0	<15			25		DISTRIB	
~10				25	Estimated true p*K* of osmic acid		
12.1				25	Distribution between CCl_4 and water	DISTRIB	Y19
7.24				20	*I* = 0.25 (Na_2SO_4)	E3bg	B30
7.2	12.2	13.95	14.17	20	Stepwise p*K* values, dil.	O	
160. Osmium tetroxide, see Osmic acid							
161. α-Oxyhyponitrous acid, $H_2N_2O_3$							
2.51	9.70			1	For *I* = 0; from *I* = 0.02 to 1.0	E3bg	S116
162. (Aquo) Palladium(II) ion, Pd^{2+}							
13.0	12.8			25	Stepwise pK_b values for Pd^{2+}	E3bg	I31
12.4	14.1			25		O6	
163. Paramolybdic acid, see Heptamolybdic acid							
164. Pentacyanoaquoferrate(II) ion, $Fe(CN)_5(H_2O)^{3-}$							
2.63				25	*I* = 0.1($LiClO_4$)	KIN	M25
165. Pentacyanoaquaferrate(III) ion, $Fe(CN)_5(H_2O)^{2-}$							
7.95				10	p*K* for $Fe(CN)_5(H_2O)^{2-} \rightleftharpoons Fe(CN)_5(OH)^{3-} + H^+$	KIN	M2
7.86				25	(but see D19)		
7.81				30			
166. Perboric acid							
7.91				25	Apparent p*K* for $H_3BO_3 + H_2O_2 \rightleftharpoons H^+ + (H_2BO_3,H_2O_2)^-$;	E3cg	A35

		$\underline{I} = 0$; from results in 0.1 M KCl		
7.71	25	Apparent $p\underline{K}$ for $H_3BO_3 + 2H_2O_2 \rightleftharpoons H^+ + (H_2BO_3,2H_2O_2)^-$; $\underline{I} = 0$		
<8	0	c = 0.02-0.07 M borate	DISTRIB	M57
7.77	18			

167. Perchloric acid, $HClO_4$

		From Raman spectroscopy, the dissociation constant is greater than 38		H55
		Perchloric acid is completely dissociated up to 6-8 M		A22
		From Raman spectroscopy, perchloric acid is completely dissociated up to 10 M		C57
−2.4 to −3.1	25		VAP	K7
−2.12	25	Assuming species is $HClO_4,7H_2O$	RAMAN, NMR	H77
−4.8	25	Assuming species is $HClO_4.3H_2O$		
−1.61	10	Molal scale; vapour pressure and isopiestic measurements; 0.1-14 mol kg^{-1}	VAP	H2
−1.54				
−1.47	40			
−1.70	0		NMR	H88
−1.58	25			
−1.42	70			
−1.63	0		NMR	K42
−1.65	25			
−7		Theoretical prediction (Ricci's method)		G37
−8		Theoretical prediction (Pauling's method)		
−7.3		Theoretical prediction		K77
−8.6		Theoretical prediction		S48
−14		$p\underline{K}$ of $H2ClO_4^+$; theoretical prediction		
		Other measurements B50, B160, E29, H87, W43		
		For Hammett acidity function of $HClO_4$, see B93, D33		

Name, Formula and pK values			T(°C)	Remarks	Methods	Reference
				(in presence of $LiClO_4$ and $NaClO_4$) H20, H27, P20, P32, Y13		
				For H_R(Jo) acidity function of $HClO_4$ see D40		
				For H_R^* acidity function of $HClO_4$ see Y9		
				For H_- acidity function of $HClO_4$ see B100		
				For H_1 acidity function of $HClO_4$ see H73		
				For H_o acidity function of $HClO_4$, see V16		
				For H_o' acidity function of $HClO_4$ in 50% ethanol, see R48, V14		
168. Perchloryl amide, $H2NClO_3$						
3.7	8.6			Titration of 0.015 M dipotassium salt with 0.1 N mineral acids	E3bg	M27
5.54	11.95			$\underline{I} = 0$	E,g	R37
5.52	11.96				E3bg	M27
169. Perchromic acid, H_2CrO_5						
4.30			22	Also $\underline{K} = 1.4$ for $H_2CrO_4 + H_2O_2 \rightleftharpoons H_2CrO_5 + H_2O$	E3ag	F32
4.95			22	Also $\underline{K} = 0.73$ for $H_2CrO_4 + H_2O_2 \rightleftharpoons H_2CrO_5 + H_2O$	O5	
170. Perhydroxyl radical, see Hydroperoxy radical						
171. Periodic acid, H_5IO_6						
1.64	8.36	14.98	25	$c = 10^{-4}$ M periodate	O5	C61
1.55	8.27		25	$\underline{I} = 0.006$ to 0.01; $c = 0.005$–0.008 M periodate; Debye-Hückel equation for extrapolation	E3bg	N12
2.23	8.01		10	$\underline{I} = 0.2$ to 1.3 ($NaNO_3$); extrapolated against $\underline{I}$	E,quin	I27
2.21	8.01		20			
2.20	8.02		30			

2.21	8.04		40			
2.22	8.07		50			
	8.34		0	c = 0.002-0.05 M periodate;	E3b	B142
	8.33		25	taking $\underline{K}$ = 2400 for dimerization constant of periodate dianion		
	8.43		45			
1.64	8.20	12.10	25	also $2H_3IO_6^{2-} \rightleftharpoons I_2O_9^{4-} + 3H_2O$, log $\underline{K}$ = 2.09	E3b	H130
1.61					O	
1.73					C	
1.61			25	Extrapolated to $\underline{I}$ = 0	KIN	S9
	8.25		25	Extrapolated to $\underline{I}$ = 0	E3bg	
				Ref. S6 gives p$\underline{K}$ values in H_2O/D_2O mixtures		
	8.10	12.0	1	$\underline{I}$ = 0.1(KCl); $c = 5.10^{-5}$–5.10^{-4} M		B141
	8.31	11.6	25			
	8.45	11.6	45			
	8.65		70			
		12.5	16	Titration of 0.2 M $K_2H_3IO_2$	E3bg	S98
~–0.8				p$\underline{K}$ for $H_5IO_6 + H^+ \rightleftharpoons I(OH)_6^+$	O6	M79
–1.0				value predicted by Ricci's method		
				True p$\underline{K}$ values of H_5IO_6 have been calculated from experimental results by assuming $\underline{K}$ = 40 for $H_4IO_6^- \rightleftharpoons 2H_2O + IO_4^-$		C59
				Other measurements: C59, L14, M58, R47		
				For H_0 acidity function of aqueous periodic acid see D29		

172. Permanganic acid, $HMnO_4$

–2.25			25	In $HClO_4$ solutions	O6	B5

Name, Formula and pK values			T(°C)	Remarks	Methods	Reference
173. Peroxydideuteriomonosulphuric acid, D_2SO_5						
10.40			19	In D_2O		K111
174. Peroxydiphosphoric acid, $H_4P_2O_3$						
	5.18	7.67	25	I = 0.01 to 1; extrapolated against $I^{1/2}$, NMe_4^+ salt titrated with HCl	E3bg	C62
−0.3	0.5			estimates, by analogy with similar acids		
175. Peroxymonophosphoric acid, H_3PO_5						
1.1	5.5	12.8	25	"Practical" constants; I = ~0.2 (for K_1) 0.14 (for K_2), ~0.15 (for K_3)	O5	B25
<1.3	~4.85		25	I = 1.5	KIN	F37
		12.5	35.8		KIN	K78
176. Peroxymonosulphuric acid, H_2SO_5						
1.0			25		ANALYT	M90
0.7			75			
	9.3		25	I ~0.2	E3bg	G55
	9.4		25	I >1; poor endpoint; decomposition	E3bg	B11
	9.86		19	I ->0	E3bg	K111
177. Peroxynitrous acid, HO.ONO						
<6			23			Y3
178. Perrhenic acid, $HReO_4$						
−1.25			25	In $HClO_4$ solutions	O6	B5
179. Pertechnetic acid, $HTcO_4$						
0.3			25.4			R52

−1.5				Predicted from pK values of $HReO_4$(−1.25) and $HMnO_4$ (−2.25)		C37
180. **Perthiocarbonic acid** (dithiodisulphidocarbonic acid), H_2CS_4						
	7.87		0	pK_2, corrected to $I = 0$	E3bg	G18
	7.54		10			
	7.41		15			
	7.32		20			
	7.24		25			
3.54			2	pK_1, corrected to $I = 0$	C1	G18
181. **Pervanadyl ion**, VO_2^+						
3.70			15	log K for $HVO_3 + H^+ \rightleftharpoons VO_2^+ + H_2O$; I = 1.0(Na,H)ClO_4	DISTRIB	Y7
3.70			20			
3.65			25			
3.80			15	−log K for $HVO_3 \rightleftharpoons VO_3^- + H^+$		
3.80			25			
3.75			35			
182. **Perxenic acid**, H_4XeO_6						
~2	~6			Estimate		A43
		10.5	24	c = 0.003 M	E3bg	
		~10			O5	
183. **Phosphine**, PH_3						
29			27	pK for $PH_3 + H_2O \rightleftharpoons PH_2^- + H_2O^+$; isotope exchange	KIN	W22
28				pK_b for $PH_4^+ + OH^- \rightleftharpoons PH_3 + H_2O$		
184. Phosphoramidic acid, see Amidophosphoric acid						

Name, Formula and pK values			T(^{o}C)	Remarks	Methods	Reference
185. Phosphoric acid, H_3PO_4						
2.056			0	$\underline{I}$ = 0.01 to 0.31; extrapolated to $\underline{I}$ = 0	Elch	B18
2.073			5			
2.088			10			
2.107			15			
2.127			20			
2.148			25			
2.171			30			
2.196			35			
2.224			40			
2.251			45			
2.277			50			
2.308			55			
2.338			60			
				$p\underline{K}_1 = 799.31/\underline{T} - 4.5535 + 0.013486\underline{T}$ ($\underline{T}$ in ^{o}K)		
				Thermodynamic values are derived from the results		
2.120	7.227	12.465	18	For $\underline{I}$ = 0	E3ah	B82
2.161	7.207	12.325	25			
2.232	7.105	12.180	37			
2.048			0.3	$\underline{I}$ = 0.01 to 0.10; extrapolated using Debye-Hückel	Elch	N41
2.076			12.5	equation		
2.124			25			
2.185			37.5			
2.260			50			
2.172			25			H58
2.126			25	For $\underline{I}$ = 0	Cl	M43
2.128			25		C	M6
1.983	7.207		20	For $\underline{I}$ = 0	E3ah	S50
2.12			25	For $\underline{I}$ = 0	CALORIM	I14

2.15		25	c = 0.0058 M; 1 atmosphere	C1	E19
2.01			500 atmosphere		
1.88			1000 atmosphere		
1.77			1500 atmosphere		
1.58			2000 atmosphere		
	7.3131	0	$\underline{I}$ = 0.02 to 0.45; extrapolated using Debye-Hückel equation;	E1a	B19,B20
	7.2817	5	c = 0.003-0.09 M (NaH_2PO_4 or KH_2PO_4),		
	7.2537	10	0.003-0.06 M (Na_2HPO_4);		
	7.2312	15	0.003-0.09 M (NaCl)		
	7.2130	20			
	7.1976	25			
	7.1891	30			
	7.1850	35			
	7.1809	40			
	7.1809	45			
	7.1831	50			
	7.1870	55			
	7.1944	60			
			Between 0° and 50°, $p\underline{K}_2 = 2073.0/\underline{T} - 5.9884 + 0.020912\underline{T}$ ($\underline{T}$ in ^oK)		
			Thermodynamic values are derived from the results		
	7.2178	20	$\underline{I}$ = 0.013 to 0.166, extrapolated to $\underline{I}$ = 0;	E1a	N40
	7.2058	25	c = 0.03 and 0.06 M Na_2HPO_4,		
	7.1973	30	0.04 and 0.06 M NaH_2PO_4, 0.03 and 0.06 M NaCl		
	7.1918	35			
	7.1890	40			
	7.1888	45			
	7.1912	50			
	7.2797	5	$\underline{I}$ = 0.024 to 0.108; extrapolated to $\underline{I}$ = 0;	E1a	G67
	7.2525	10	c = 0.006-0.026 M KH_2PO_4,		
	7.2305	15	0.003-0.013 M $KNaHPO_4$, 0.010-0.043 M NaCl		

Name, Formula and pK values	T(^{o}C)	Remarks	Methods	Reference
7.2129	20			
7.2004	25			
7.1902	30			
7.1828	35			
7.1783	40			
7.1758	45			
7.1764	50			
		$p\underline{K}_2 = 1775.812/\underline{T} - 3.9762 + 0.0175089\underline{T}$ ($\underline{T}$ in ^{o}K)		
		Thermodynamic values are derived from the results		
7.1988	25	$\underline{I}$ = 0.03 to 0.25; extrapolated to $\underline{I}$ = 0	Ela	E28
7.1891	30			
7.1814	35			
7.1777	40			
7.1776	45			
7.1796	50			
7.1859	55			
7.1918	60			
		Results are also given for 10% and 20% methanol-water solutions		
		Thermodynamic values are derived from the results		
7.21	60		E,g	C68
7.24	70			
7.24	80			
7.24	90			
12.375	25	$\underline{I}$ = 0.01 to 0.51; extrapolated to $\underline{I}$ = 0; taking $p\underline{K}$ = 10.329 for HCO_3^-	O3	V4
		For values of $p\underline{K}_1$ in H_2O/D_2O mixtures, see S3		
		For values of $p\underline{K}_2$ in H_2O/D_2O mixtures, see R51		
		On the assumption that some H_3PO_4 dimerizes to $H_6P_2O_4$		E26

				in strong solutions, a pK_1 value of 0.52 is derived for the latter from conductance measurements		
		12.043	10	pK_3; I = 0.03 to 0.10, extrapolated to I = 0	E3ah	G33
		12.005	15			
		11.979	20			
		11.962	25			
		11.959	30			
		11.957	35			
		11.957	40			
		11.977	45			
		11.971	50			
	7.631	12.884	0	pK_2, pK_3; I = 0.1-1.0, extrapolated to I = 0	E2bh	M62
	6.796	11.848	25			
	6.079	10.983	50			
	5.464	10.259	75			
	4.933	9.647	100			
	4.471	9.127	125			
	4.068	8.681	150			
	3.710	8.294	175			
	3.390	7.954	200			
	3.101		225			
	2.835		250			
	2.588		275			
	2.355		300			
1.713	6.730		0.2	pK_1, pK_2; I = 0.5, extrapolated to I = 0	E3bg	P23
1.742	6.712		10			
1.860	6.690		22			
1.987	6.715		30			
2.008	6.718		40			
2.108	6.726		50			

Name, Formula and pK values			T(°C)	Remarks	Methods	Reference
				$pK_1 = 583.01/T - 2.715 + 0.00980/T$ (T in °K) for range 25-175°C		P22
				$pK_2 = 1272/T - 1.154 + 0.01368T$ $pK_1 = 2.150 - AI^{1/2}(1 + B(1.66)/I^{1/2})^{-1}$ Effect of ionic strength on pK_1 at 25°		K67
1.89	6.270	10.85		$I = 3(NaClO_4)$	E,g,h	B9
	6.14	10.44	15	"Apparent" constants in artificial sea-water ($I = 0.689$)		C17
				For pK_a values in organic solvents, see K90, L58, T21		
0.09			25	pK in conc. H_2SO_4	Cl	F22
1.95	6.70	11.30	37	"Practical" constants; 0.002-0.012 M in H_3PO_4; $I = 0.15(KNO_3)$ Also $2H_2PO_4^- \rightleftharpoons H_4(PO_4)_2^{2-}$, log K = 0.75 $HPO_4^{2-} + H_2PO_4^- \rightleftharpoons H_3(PO_4)_2^{3-}$, log K = 0.43	E3bg	C27
	7.211		25	pK_2; extrapolated to $I = 0$	E3	M21
		11.921	25	pK_3; $I = 0$ Also $2H_2PO_4^- \rightleftharpoons H_4P_2O_8^{2-}$, log K = 0.22 $2HPO_4^{2-} \rightleftharpoons H_2P_2O_8^{4-}$, log K = −0.52	RAMAN	P60
				Log K = 2.16 for $H_2PO_4^- + H_2PO_4^- + H^+ \rightleftharpoons H_5(PO_4)^{5-}$; at 25° and $I = 3(NaCl)$ Log K = -0.14 for $2H_2PO_4 \rightleftharpoons H_4(PO_4)_2^{2-}$ Log K = 12.24 for $2HPO_4^{2-} + 2H^+ \rightleftharpoons H_4(PO_4)_2^{2-}$ Log K = 6.69 for $2HPO_4^{2-} + H^+ \rightleftharpoons H_3(PO_4)_2^{3-}$	E3bg	I24
1.98	6.77	11.28	25	Stepwise pK values; $I = 3(KCl)$; $c = 0.05-1.00M$ Also 0.39 = −log K for $2H_3PO_4 \rightleftharpoons (H_3PO_4)_2$ (= $H_6P_2O_8$) 1.62 = pK of $H_6P_2O_8$ 2.56 = pK of $H_2P_2O_8$ 0.38 = −log K for $2H_2PO_4^- \rightleftharpoons H_4P_2O_8^{2-}$	E3bg	F19

				$5.98 = p\underline{K}$ of $H_4P_2O_8^{2-}$		
				$8.32 = p\underline{K}$ of $H_3P_2O_8^{3-}$		
				$0.77 = -\log \underline{K}$ for $2HPO_4^{2-} \rightleftharpoons H_2P_2O_8^{4-}$		
				For the effect of $[Na^+]$ on dissociation of H_3PO_4, see M54		
				For the effect of pressure on $p\underline{K}$, see N30		
				Other measurements: A1, B83, B116, B122, B132, B135, C25, C39, D1, D61, E15, F20, F42, G1, G65, G70, H11, I13, J22, J23, K57, K72a, K101, L9, L59, L60, M14, M23, M62, M70, M71, M92, M99, P61, P63, R51, S51, S59, S69, S79, T8, W17		
				For Hammett acidity function of H_3PO_4, see D57, G27, G28 (temperature range), H54, L17 (in the presence of $NaClO_4$), P20		
				For H_0', H_0''', H_R and H_R' acidity functions of H_3PO_4, see A51		
186. Phosphoric triamide, $PO(NH_2)_3$						
<3.6			25	c = 0.1 M	E	F11
187. Phosphorous acid, H_3PO_3						
1.94	6.73			c = 0.01-0.04M; extrapolated to $\underline{I} = 0$		T3
~1.8	6.70		18	For $\underline{I} = 0$	E,quin	K57
1.20	6.70	>14	20	For $\underline{I} = 0$; measurements at three concentrations were assumed to fit a curve, $p\underline{K} = p\underline{K}_0 + a\underline{I}^b$	E3ah	F45
−5.0			20	c = 1M H_3PO_3 in H_2SO_4; phenolphthalein cation as indicator	O6	
				Other measurements: B83, G65, M99, N48		
				For Hammett acidity function of H_3PO_3, see B16		
	6.79		25	$\underline{I}$ = 0.1-1.0, extrapolated to $\underline{I} = 0$	E3bg	M22
1.43			25	Extrapolated to $\underline{I} = 0$	KIN	S6
	6.54		25	Extrapolated to $\underline{I} = 0$	E3bg	

Name, formula and pK values	T(^{o}C)	Remarks	Methods	Reference
		For pK values in organic solvents, see K90		
188. (Aquo) Plutonium(III) ion, Pu^{3+}				
7.22	25	pK for hydrolysis of Pu^{3+}; I = 0.07 ($HClO_4$)	E3bg	K85
7.37	25	I = 0.02 (HCl)		
6.95		I = 0.05	DISTRIB	K82
7.30		I = 0.02-0.07		M102
7.12				M51
189. (Aquo) Plutonium(IV) ion, Pu^{4+}				
1.77	0	pK for hydrolysis of Pu^{4+}; I = 2($NaClO_4$)		R4
1.51	12.5			
1.27	25			
1.41	15	pK for hydrolysis of Pu^{4+}; I = 2($LiClO_4$, $HClO_4$)	REDOX	R3
1.26	25			
1.06	34.4			
1.51	25	I = 1($NaClO_4$)	REDOX	R7
1.60	25	I = 0.5($NaClO_4$)	O6	K85
0.70	25	For I = 0		
1.9	15.4	I = 2($NaClO_4$)	O6	R6
1.73	25			
1.94	25	In D_2O, I = 2($NaClO_4$)		
1.6	25	I = 1.1(NaCl)	O6	H71
1.05	25	In D_2O; I = 1($NaClO_4$)	KIN	R5
0.45 0.75 3.3 6.3		Stepwise pK values for hydrolysis of Pu^{4+}; I = 1 ($LiClO_4$,$HClO_4$) [Pu^{4+}] from 10^{-7}–10^{-8} M; from [H^+] = 1M to pH 7.5	DISTRIB	M58,M66, M67
1.54 1.70		Stepwise pK values 5.10^{-6} M in Pu^{4+}	ION,D	D9

190. (Aquo) Plutonium(V) ion, PuO_2^+							
9.6				25	p*K* for hydrolysis of PuO_2^+; *I* = 0.003 (HCl)	E3b	K82
9.7					*I* = 0.003 ($HClO_4$)		
191. (Aquo) Plutonyl ion, PuO_x^{2+}							
3.39	5.25	9.52		20	Successive p*K* values for hydrolysis of PuO_2^{2+}; $(PuO_2)_2(OH)_3^+$ and $(PuO_2)_2(OH)^+$; are also formed	SOLY	M101
3.33	4.05				p*K* values for hydrolysis of PuO_2^{2+} in dilute HNO_3	E3ag	K91
5.71	5.71	9.7		25	p*K* values for hydrolysis of PuO_2^{2+}, ignoring polynuclear complexes	E,g	K81
8.21				25	–log *K* for $2PuO_2^{2+} + 2H_2O \rightleftharpoons (PuO_2)_2(OH)_2^{2+} + 2H^+$; *I* = 3($NaClO_4$) *c* = 10-150 mM PuO_2^{2+} Higher complexes are also found	E3bg	S25,S26
5.97				25	–log *K* for $PuO_2^{2+} + H_2O \rightleftharpoons (PuO_2)(OH)^+ + H^+$; *I* = 1($NaClO_4$); *c* = 0.3-30 mM in PuO_2^{2+}	E	C12
8.51					–log *K* for $2PuO_2^{2+} + 2H_2O \rightleftharpoons (PuO_2)_2(OH)_2^{2+} + 2H^+$		
22.16					–log *K* for $3PuO_2^{2+} + 5H_2O \rightleftharpoons (PuO_2)_2(OH)_5^+ + 5H^+$		
					Other measurements: M111		
192. (Aquo) Polonium(IV) ion, Po^{4+}							
0.48	2.74	5.58	9.0		Stepwise p*K* values for hydrolysis of Po^{4+}; *I* = 1($NaClO_4$); pH 0.5-3.2	DISTRIB	A29
193. Polymolybolic acid							
7.92					–log *K* for $H_2MoO_4 \rightleftharpoons 2H^+ + MoO_4^{2-}$; pH = 3.0-5.4; [$MoO_4^{2-}$] = 3-13 mM	E,g	B10
48.93					–log *K* for $H_8(MoO_4)_6^{4-} \rightleftharpoons 8H^+ + 6MoO_2^{2-}$		
59.12					–log *K* for $H_9(MoO_4)_6^{3-} \rightleftharpoons 9H^+ + 6MoO_4^{2-}$		
194. Polyphosphoric acid, $H_{62}P_{60}O_{181}$.					See also Hexadeca-, Hexa-, Tri- and Tetra-polyphosphoric acids		
7.22	8.17			25	Concentration constants; *I* = 1(NMe_4Br); *f*± assumed same as	E3bg	I13

Name, formula and pK values	T(^{o}C)	Remarks	Methods	Reference
		for HBr		
7.28 8.03	37	Concentration constants, as above	E3bg	I12
7.28 8.03	50			
195. (Aquo) Potassium ion, K^{+}				
−2.0 to −2.5	25	pK_b For pK values of KOH in superheated steam between 400 and 700°, with densities from 0.3 to 0.8 g/cm^3, see F40 For H- acidity function of KOH see S53, Y1, Y2	VAP	K7
196. (Aquo) Praseodymium(III) ion, Pr^{3+}				
8.5	25	pK for hydrolysis of Pr^{3+} to $PrOH^{2+}$; I = 3($NaClO_4$)	E3b,quin	T13
8.55	25	pK_a for hydrolysis of Pr^{3+}; titration of 0.004-0.009 M $Pr(ClO_4)_3$ with 0.02 M $Ba(OH)_2$; I = 0.3($NaClO_4$)	E3b	F49
~9	25	pK for hydrolysis of Pr^{3+}; hydrolysis of "pure" salts; c = 0.001-0.025M $Pr_2(SO_4)_3$	E3ag	M87
197. (Aquo) Protoactinium(IV) ion, Pa^{4+}				
0.14 0.38 1.25		Successive pK values for hydrolysis to $PaOH^{3+}$, $Pa(OH)_2^{2+}$ and $Pa(OH)_3^{+}$; I = 3($HClO_4$, $LiClO_4$)	DISTRIB	G72
198. (Aquo) Protoactinium(V) ion, Pa^{5+}				
1.05	25	pK for $Pa(OH)_3^{2+} + H_2O \rightleftharpoons Pa(OH)_4^{+} + H^{+}$; I = 3($LiClO_4$ $HClO_4$)	DISTRIB	G71
4.50		−log K for $PaO(OH)_2^{+} + 2H_2O \rightleftharpoons Pa(OH)_5 + H^{+}$; I = 3($NaClO_4$)	DISTRIB	G73
199. Pyrophosphoric acid, $H_4P_2O_7$				
2.28 6.70 9.37	25	For I = 0	E3bg	N13
9.53	25	I = 0.005 to 0.035; extrapolated to I = 0	E3bh	W32

			9.57	40			
		6.68	9.39	18	For $\underline{I}$ = 0, calculated from measured pH values of pairs of salts on progressive dilution; corrected for hydrolysis	E3ah	K60
		6.57	9.62	25	For $\underline{I}$ = 0; $\underline{I}$ = 0.0015 to 0.0019 (for $\underline{K}_3$) $\underline{I}$ = 0.004 to 0.018 (for $\underline{K}_4$); activity coefficients calculated by Debye-Hückel equation	E3bg	M91
		6.70	9.88	30	For $\underline{I}$ = 0; salts added to sodium pyrophosphate, HCl mixtures to vary $\underline{I}$ (0.0015 to 0.24 for $\underline{K}_3$, 0.021 to 0.34 for $\underline{K}_4$); extrapolation using Debye-Hückel equation	E3ah	M98
1.52	2.36	6.60	9.25	25	For $\underline{I}$ = 0	E3bg	D26
0.44				25	For $\underline{I}$ = 0	CALORIM	I14
	2.27	6.63	9.29	25	For $\underline{I}$ = 0; extrapolated from results for $\underline{I}$ ~0.01 (K^+ salt), using Debye-Hückel equation	E3bg	B53
	2.64	6.76	9.42	25	NMe_4^+ salt solutions; extrapolation to $\underline{I}$ = 0 against (concentration)$^{1/2}$; $P_2O_7^{4-}$ forms complexes with K^+, Na^+ and Li^+	E3bg	L7
	2.22	6.36	9.11	25	"Practical" constants; $\underline{I}$ = 0.1(NMe_4Cl);		
0.82	1.81	6.13	8.93	25	"Practical" constants; $\underline{I}$ = 1(NMe_4Cl); p$\underline{K}_1$ value is experimentally uncertain		
2.3	2.5	6.17	9.08	0	Concentration constants; $f\pm$ assumed same as for HBr; $\underline{I}$ = 0.1 (NMe_4Br); the p$\underline{K}_1$ values are uncertain	E3bg	I12
2.2	2.3	6.03	8.97	10			
2.0	2.0	6.12	8.95	25	$\underline{I}$ = 0.1		
1.9	1.95	6.13	8.94	37	$\underline{I}$ = 0.1		
1.7	1.97	6.12	8.93	37	$\underline{I}$ = 0.2		
1.7	1.91	6.08	8.88	37	$\underline{I}$ = 0.3		
1.9	1.98	6.13	8.97	50	$\underline{I}$ = 0.1		
1.3	1.92	6.06	8.90	50	$\underline{I}$ = 0.2		
1.2	2.12	6.04	8.88	50	$\underline{I}$ = 0.3		
1.3	2.12	6.16	8.92	65	$\underline{I}$ = 0.1		
1.2	2.17	6.01	8.72	65	$\underline{I}$ = 1.0		

Name, Formula and pK value				T(^{o}C)	Remarks	Methods	Reference
0.88	2.00	6.28	9.10	25	Probably "practical" constants; NMe_4^+ pyrophosphate solutions; $\underline{I}$ = 0.05(NMe_4Br); values of $p\underline{K}_1$ are experimentally uncertain	E3bg	M83
0.84	1.96	6.12	9.01		$\underline{I}$ = 0.10		
0.82	1.83	6.05	8.87		$\underline{I}$ = 0.42		
0.79	1.72	5.76	8.71		$\underline{I}$ = 1.00		
1.04	2.04	6.37	9.18	50	$\underline{I}$ = 0.05		
0.95	1.99	6.34	9.11		$\underline{I}$ = 0.10		
0.86	1.82	5.90	8.77		$\underline{I}$ = 0.42		
0.83	1.64	5.74	8.64		$\underline{I}$ = 1.14		
1.14	2.55	6.38	9.26	60	$\underline{I}$ = 0.05		
1.05	2.02	6.33	9.19		$\underline{I}$ = 0.10		
0.98	1.76	5.94	8.79		$\underline{I}$ = 0.42		
0.98	1.60	5.72	8.62		$\underline{I}$ = 1.14		
1.04	1.97	6.26	9.23	65	$\underline{I}$ = 0.05		
0.97	1.94	6.26	9.16		$\underline{I}$ = 0.10		
0.92	1.71	5.90	8.77		$\underline{I}$ = 0.42		
0.91	1.54	5.72	8.61		$\underline{I}$ = 1.14		
1.00	1.91	6.17	9.16	70	$\underline{I}$ = 0.05		
0.94	1.89	6.14	9.06		$\underline{I}$ = 0.10		
0.89	1.66	5.87	8.71		$\underline{I}$ = 0.42		
0.97	1.50	5.72	8.58		$\underline{I}$ = 1.14		
0.97	2.12	5.84	8.01	65.5	"Practical" constants; c = 0.08-0.18 M pyrophosphate	E3b	M8
		5.68	8.00	27.4	"Practical" constants; $\underline{I}$ = 0.75($NaNO_3$)	E3bg	Y8
2.5	2.7	6.0	8.3	25	Probably "practical" constants; $\underline{I}$ = 0.1($NaNO_3$)	E3bg	J8
~1.7	1.75	5.98	8.74	25	Concentration constants; $\underline{I}$ = 1(NMe_4Br); NMe_4^+ pyrophosphate solutions; $f\pm$ assumed same as for HBr	E3bg	I13
	2.52	6.08	8.45	20	"Practical" constants; $\underline{I}$ = 0.1(KCl)	E3bh	S54
		5.61	7.68	25	"Practical" constants; $\underline{I}$ = 1(KNO_3)	E3bg	W10

0.70	2.19	6.80	9.59				E7
1.38	1.88	5.97	8.46	25	I = 0.5(NMe_4Cl)	E	D34
		6.12	8.93	25	I = 0.1(NMe_4Cl)	E3bg	M13
			7.45	25	I = 1($NaNO_3$)	E3bg	C47
			6.96	25	I = 1($NaClO_4$)		
0.91	2.10	6.70	9.32	25	$I \to 0$	E3bg	M83
					Other measurements: A1, F40, K33, K51, M99, M112, O22		

200. Pyrosulphuric acid, $H_2S_2O_7$

3.1	25	pK in CF_3COOH	C	F25

201. Rhenic acid, H_5ReO_6

1.63	7.69	pK_1, pK_2 for H_5ReO_6	C, E, O	S82

202. (Aquo) Rhodium(III) ion, Rh^{3+}

3.43		pK for $Rh^{3+} \rightleftharpoons RhOH^{2+} + H^+$; c = 0.0015M; polymerized species probably form slowly	E3bg	F35
3.2	25	pK for Rh^{3+}	E3bg	P50
2.4	64.4		EST	
2.92	20	I = 1($NaClO_4$); bridged complexes probably form slowly	O5	C41
3.40	25	pK for hydrolysis of Rh^{3+}; hydrolysis of "pure" salts	E3ag	S128
3.20	45			
3.08	60			

203. Ruthenium tetroxide, see Diperruthenic acid

204. (Aquo) Samarium(III) ion, Sm^{3+}

8.34	25	pK_a for hydrolysis of Sm^{3+}; titration of 0.004-0.009 M $Sm(ClO_4)_3$ with 0.02 M $Ba(OH)_2$; I = 0.3($NaClO_4$)	E3b	F49

Name, Formula and pK value			T (^{o}C)	Remarks	Methods	Reference
7.5			25	–log K for $Sm^{3+} + H_2O \rightleftharpoons SmOH^{2+} + H^+$; I = 1		K78
15.0				–log K for $Sm^{3+} + 2H_2O \rightleftharpoons Sm(OH)_2{}^+ + 2H^+$		
22.7				–log K for $Sm^{3+} + 3H_2O \rightleftharpoons Sm(OH)_3 + 3H^+$		
19.5				–log K for $3Sm^{3+} + 4H_2O \rightleftharpoons Sm_3(OH)_4{}^- + 4H^+$		
205. (Aquo) Scandium(III) ion, Sc^{3+}						
4.93			25	pK for hydrolysis of Sc^{3+}; I = 1($NaClO_4$); c = 0.001-0.02M $Sc(ClO_4)_3$; also log K = 3.87 for $2ScOH^{2+} \rightleftharpoons Sc_2(OH)_2{}^{4+}$	E3a,quin	K30
5.09			10	pK for hydrolysis of Sc^{3+}; I = 1($NaClO_4$);	E3a,quin	K31
4.41			40	log K for dimerization is 3.53 at 10°, 3.33 at 40°		
4.61			25	I = 0.01		
5.1	5.1		25	Successive pK values for hydrolysis of Sc^{3+} to $ScOH^{2+}$ and $Sc(OH)_2{}^+$; I = 1($NaClO_4$); the main species are polynuclear species, $Sc[(OH)_2Sc]_n{}^{(3+n)+}$; from re-examination of data of M. Kilpatrick and L. Pokras, J. Electrochem. Soc. 100, 85 (1953); 101, 39 (1954)		B64
5.11			25	pK for hydrolysis of Sc^{3+}; I = 1($NaClO_4$); also –log K = 6.14 for $2Sc^{3+} + 2H_2O \rightleftharpoons Sc_2(OH)_2{}^{4+} + 2H^+$; –log K = 13.00 for $3Sc^{3+} + 4H_2O \rightleftharpoons Sc_3(OH)_4{}^{5+} + 4H^+$; –log K = 17.47 for $3Sc^{3+} + 5H_2O \rightleftharpoons Sc_3(OH)_5{}^{4+} + 5H^+$	E3bg	A59
9.08				log K for $Sc^{3+} + OH^- \rightleftharpoons ScOH^{2+}$; $c = 4.10^{-5} - 10^{-2}$M in Sc^{3+}		D13
9.29	8.66	7.89	25	successive log K values for Sc^{3+}, $ScOH^{2+}$, $Sc(OH)_2{}^+$ from Sc^{3+} and OH^- $c = 10^{-3}$M Sc^{3+}; I = 1(KNO_3, KSCN, Na_2SO_4 or $NaClO_4$)	E3bg	K69
4.90	5.78	6.58	20	successive pK values for the hydrolysis of Sc^{3+}; I = 0.1	O8	A41
4.55	8.76		25	successive pK values for the hydrolysis of Sc^{3+}; I = 0.05; c = 0.005M $Sc(ClO_4)_3$	E3bg	H2
4.74			25	pK for $Sc^{3+} + H_2O \rightleftharpoons ScOH^{2+} + H^+$; I = 0.1; $[Sc^{3+}] = 1.25.10^{-3}–2.0.10^{-2}$M		K65

4.47		25	$\underline{I}$ = 0.1(KNO_3); Sc^{3+} = 2.6.10^{-4}-2.6.10^{-3}M	O8,E3bg	A21
5.86			−log $\underline{K}$ for $2Sc^{3+} + 2H_2O \rightleftharpoons Sc_2(OH)_2^{4+} + 2H^+$		
			Other measurements: W29		
206. Selenic acid, H_2ScO_4					
	1.36	0	$\underline{I}$ = 0.007 to 0.019; extrapolated to $\underline{I}$ = 0	Elch	N8
	1.46	10	using Davies' equation		
	1.52	15			
	1.58	20			
	1.66	25			
	1.73	30			
	1.82	35			
	1.89	40			
	1.96	45			
			Thermodynamic quantities are derived from the results.		
	1.83	0	Extrapolated to $\underline{I}$ = 0, using the Debye-Hückel limiting	Elc,quin	P10
	1.845	5	expression for activity coefficients		
	1.86	10			
	1.87	15			
	1.90	20			
	1.92	25			
	1.95	30			
	1.70 to 1.78	25	Recalculation of data by V.S.K. Nair (J. Inorg. Nuclear Chem., 26, 1911 (1964)), using an extended Debye-Hückel equation; the value of $p\underline{K}_2$ is sensitive to the choice of ion size parameter		C51
			For $\underline{H}_o$ values of selenic acid, see M4, W9		
			Other measurements: G26		
207. Selenious acid, H_2SeO_3					
2.62	8.32	25	For $\underline{I}$ = 0	E,g	H6

Name, Formula and pK value		T (°C)	Remarks	Methods	Reference
2.54	8.02	18	Titration of 0.04M H_2SeO_3	E3bg	B122
2.42	8.08			E3bg	R53
2.40	8.06		c = 0.05M H_2SeO_3; added NaOH	O4	W26
2.46		25		C1	R36
2.40			0.01N solutions	O4	
2.64	8.27	25	$\underline{I}$ = 0.1(KCl); concentration constants	E3bg	K20
			Values are also given for methanol-, ethanol-, and 2-propanol-water mixtures		
2.33		25	concentration constant, $\underline{I}$ = 0.1	DISTRIB	S58
3.10	8.30	20	$\underline{I}$ = 0	E3bg	N25
2.75	8.50	25	Extrapolated to $\underline{I}$ = 0	E3bg	K20
2.62	8.25	25	Extrapolated to $\underline{I}$ = 0	E3bg	S6
			For values of p$\underline{K}$ in D_2O/H_2O mixtures, see S6		
2.61	8.05	25	$\underline{I}$ = 3($NaClO_4$)	E3bg	B13
	7.79		log $\underline{K}$ for $H^+ + 2SeO_3^{2-} \rightleftharpoons H(SeO_3)_2^{3-}$		
	15.49		log $\underline{K}$ for $2H^+$ $2SeO_3^{2-} \rightleftharpoons (HSeO_3)_2^{2-}$		
	19.02		log $\underline{K}$ for $3H^+ + 2SeO_3^{2-} \rightleftharpoons H(HSeO_3)_2^{-}$		
	20.91		log $\underline{K}$ for $4H^+ + 2SeO_3^{2-} \rightleftharpoons (H_2SeO_3)_2$		
2.70	8.24	25	$\underline{I}$ = 3(NaCl). Dimeric species are also formed	E3bg	S2
			$pK_1 = -243/T + 3.344$ ($\underline{T}$ in °K)	LIT	E8
			$pK_2 = 243/T + 5.792$		
			Other measurements: B63, G31, V6.		

208. Selenocyanic acid, HSeCN

<1	25	from hydrolysis of salts	E,g	B98

209. Silicic acid, H_4SiO_4

9.77	25	pK_1; extrapolated against $\underline{I}^{1/2}$ to $\underline{I}$ = 0	SOLY	G61

9.70		35			
9.85	11.8	20	pK_1, pK_2; calculated from published e.m.f. data;		G60
9.7	11.9	25	I = 0.02 to 0.04; extrapolated against I to I = 0		
9.1	11.9	30			
9.9		25		Cl	
9.51	11.77	25	pK_1, pK_2	E3bh	O10
9.66	11.70	30	pK_1, pK_2; $pK_3 \sim pK_4 \sim 12$	E3ah	F27
9.4	11.4	20	pK_1, pK_2; pK_3 = 13.7	E,g,h	M72
9.91		25	pK_1, H_4SiO_4 prepared by hydrolysis of its methyl ester; I = 0	Cl	S45
9.46		25	pK_1; I = 0.5($NaClO_4$)	E	B75
9.3		60	In 0.01M borax solutions; equilibrated with powdered quartz	SOLY	V5
9.1		70			
9.1		80			
9.1		90			
9.1		100			
9.2		90	In unbuffered solutions of alkali	SOLY	
8.83		346	160 atmospheres	SOLY	V64
9.32		355	180 atmospheres		
10.33		364	200 atmospheres		
10.28		0	$-\log K$ for $Si(OH)_4 \rightleftharpoons SiO(OH)_3^- + H^+$; I = 0.1-5.0; extrapolated to I = 0	E3bh	B158
			Polysilicate formation was studied in 1M NaCl at 60-290° and Si(IV) concentrations of 0.005-0.05M		
9.82		25			
9.50		50			
9.27		75			
9.10		100			
8.98		125			
8.90		150			

Name, Formula and pK value		T(°C)	Remarks	Methods	Reference
8.85		175			
8.85		200			
8.89		225			
8.96		250			
9.07		275			
9.22		300			
-	11.81	25	corrected to $\underline{I} = 0$	Cl	R64
-	10.11	50			
8.20	9.24	100			
8.46	9.10	150			
8.59	9.48	200			
8.69	9.81	250			
4.3		25	log $\underline{K}$ for $Si(OH)_4 + OH^- \rightleftharpoons SiO(OH)_3^- + H_2O$; $\underline{I}$ = 0.5 NaCl		
5.3		25	log $\underline{K}$ for $Si(OH)_4 + 2OH^- \rightleftharpoons SiO_2(OH)_2^{2-} + 2H_2O$		
15.1		25	log $\underline{K}$ for $4Si(OH)_4 + 2OH^- \rightleftharpoons Si_4O_6(OH)_6^{4-} + 6H_2O$		
			Other measurements: B87, H3, H10, H32, I3, S62 (in 0.5 M NaCl), J20, L2 (in 0.5 M $NaClO_4$ and 3M $NaClO_4$), M108 (colloidal silicic acid), R36 (colloidal silicic acid)		
210. (Aquo) <u>Silver ion</u>, Ag^+					
>11.1		25	p$\underline{K}$ for hydrolysis of Ag^+; $\underline{I}$ = 1($AgNO_3$) Hydrolysis of Ag^+ gives AgOH, $Ag(OH)_2^-$ and possibly poly-nuclear species	E,g	B63
3.50		25	log $\underline{K}$ for $Ag^+ + 2OH^- \rightleftharpoons Ag(OH)_2^-$; $\underline{I}$ = 3($NaClO_4$); no appreciable amounts of AgOH formed	DISTRIB	A38
3.60		25	$\underline{I}$ = 3($NaClO_4$)	SOLY	
3.99		25	log $\underline{K}$ for $Ag^+ + 2OH^- \rightleftharpoons Ag(OH)_2^-$	SOLY	B68
3.64		25	log $\underline{K}$ for $Ag^+ + 2OH^- \rightleftharpoons Ag(OH)_2^-$; calculated from data	SOLY	R19

		given in J12		
12.1	25	Estimated p*K* of AgOH ⇌ AgO^- + H^+	SOLY	J12
3.02	25	*I* = 1($NaClO_4$); log *K* for Ag^+ + OH^- ⇌ AgOH	SOLY	G68
4.69		log *K* for Ag^+ + $2OH^-$ ⇌ $Ag(OH)_2^-$		
		Other measurements: C10, F6, G77, K73, L29, L52		

211. (Aquo) Sodium ion, Na^+

-0.81	5	pK_b; *I* = 0.02 to 0.1; *f*± calculated using Davies'	Elch	G40
-0.81	15	equation; e.m.f. data of H.S. Harned and G.E. Mannweiler,		
-0.77	25	J. Am. Chem. Soc., 57, 1873, (1935).		
-0.88	35			
-0.81	45			
-0.45	5	pK_b; *I* = 0.02 to 0.1; *f*± calculated using Davies'	Elch	
-0.46	15	equation; e.m.f. data of H.S. Harned and W.J. Hamer,		
-0.57	25	J. Am. Chem. Soc., 55, 4496, (1933)		
-0.72	35			
-0.62	45			
~-0.7	25	pK_b	CAT,KIN	B40
-1.9	25	pK_b	VAP	K7
		For alkalinity function for NaOH solutions, see M89		
		For indicator acidity function H_- for NaOH solutions, see E4, S53, Y1, Y2		
		For J_- function for NaOH solutions see B99, G49		

212. (Aquo) Stannous ion, see (Aquo) Tin(II) ion

213. (Aquo) Strontium ion, Sr^{2+}

0.78	5	pK_b of $SrOH^+$; *I* = 0.02 to 0.1; *f*± calculated using	Elch	G40
0.80	15	Davies' equation; from e.m.f. data due to H.S. Harned		
0.82	25	and T.R. Paxton, J. Phys. Chem., 57, 531, (1953)		

Name, Formula and pK value	T(°C)	Remarks	Methods	Reference
0.86	35			
0.89	45			
		Thermodynamic quantities are derived from the results		
0.82		pK$_b$ of SrOH$^+$; concentration constant; $c = 0.2 - 1N(SrCl_2$ and $Sr(NO_3)_2)$; salt effect on indicator	O3	K55
0.23	25	pK$_b$ of SrOH$^+$; $\underline{I} = 3(NaClO_4)$	E2ah	C7
0.96	25	$\underline{I} = 0.02$ to 0.065; extrapolation to $\underline{I} = 0$ using Davies' equation; solubility of $Sr(IO_3)_2$ in NaOH solutions	SOLY	C43

214. <u>Sulphamic acid</u>, NH$_2$SO$_3$H

Name, Formula and pK value	T(°C)	Remarks	Methods	Reference
1.03	10	For $\underline{I} = 0$; in most of the cells, concentration of	E1a	K34
1.02	15	NH$_2$SO$_3$H, NH$_2$SO$_3$Na and NaCl were approximately the		
0.99	20	same (0.005–0.054M)		
0.99	25			
0.98	30			
0.98	35			
1.00	40			
1.025	45			
1.04	50			
0.979–1.013	25	For $\underline{I} = 0$; value varies slightly with method of calculation	C1	S102
1.00	25	For $\underline{I} = 0$	C1,R1d	T7
0.58	95	$\underline{I} = 1(NaClO_4)$	KIN	C2
1.19	20	dilute solutions	E3bg	H30
		This paper also gives p$\underline{K}$ values in aqueous mixed solvents		
1.055	25	$\underline{I} = 0$	E1ag	M92

215. <u>Sulphotitanic acid</u>, TiO(OH)OSO$_2$OH

Name, Formula and pK value	T(°C)	Remarks	Methods	Reference
4.40	25	$\underline{I} = 0.5$–8.0	E,g	G57

216. Sulphuric acid, H_2SO_4

1.58	0	pK_2; I = 0.005 to 0.02; $\gamma\pm$ calculated from extended Debye-Hückel equation; values of pK_2 are sensitive to the parameters used in the Debye-Hückel equation	Elch	N9
1.63	5			
1.80	15			
1.96	25			
2.09	35			
2.22	45			
1.91	18	pK_2; for I = 0; re-examination of literature values obtained from conductivity data		K26
1.99	25			
2.28	50			
1.76	5	pK_2; I = 0.01 to 0.04; extrapolated using extended Debye-Hückel equation	Elch	D22
1.80	10			
1.84	15			
1.92	20			
1.99	25			
2.05	30			
2.11	35			
2.17	40			
2.30	50			
1.66	0	I = 0; from published conductivity data, using limiting Onsager equation		
1.68	0	I = 0; from published freezing point depression data	FP	
1.90	10	pK_2; for I = 0	E2bg	K32
2.00	20			
2.04	25			
2.13	30			
2.17	35			
2.22	40			
1.989	25	pK_2; I = 0, molal scale	Elch	M92
2.148	25	pK_2; I = 0, molal scale	Elb	G69

Name, Formula and pK value	T(^oC)	Remarks	Methods	Reference
	25	$p\underline{K}_2 = 1.95 - 2.04\ \underline{I}^{1/2}/(1 + 0.85\ \underline{I}^{1/2})$ up to $\underline{I} = 2.5$	RAMAN	T31
		For values of $p\underline{K}_2$ in D_2O/H_2O mixtures, see S6		
1.987	25	For $\underline{I} = 0$; re-examination of literature values from conductometry, spectrophotometry and potentiometry		D69
1.975	25	For $\underline{I} = 0$; Ag/AgCl electrode replaced by Hg/Hg_2SO_4; ion parameter in Debye-Hückel equation taken as 1.9	Elch	C53
1.94	25	For $\underline{I} = 0$; Ag_2SO_4 in $NaClO_4/HClO_4$ solutions	SOLY	K24
1.99	25		RAMAN	Y20
1.89	25	$p\underline{K}_2$; $\underline{I} = 0$; molal scale; solubility of Ag_2SO_4 plotted as function of ionic strength	SOLY	L40
2.37	50			
2.70	75			
3.01	100			
3.33	125			
3.69	150			
4.09	175			
4.49	200			
4.94	225			
		$p\underline{K}_2 = 1283.108/\underline{T} - 12.31995 + 0.04223215\ \underline{T}$ ($\underline{T}$ in ^oK) Thermodynamic values are derived from the results		
2.60	100	$p\underline{K}_2$; molar scale; at density of 1 $g.cm^{-3}$; values at lower densities are also given	Cl	Q7
2.83	150			
3.13	200			
3.35	250			
3.58	300			
3.08	100	Molal scale	Cl	
4.03	200			
1.88	25	$p\underline{K}_2$; for $\underline{I} = 0$; using cation-permselective membrane	ION	W2
2.03	35			
2.21	50			

1.99	25	pK_2; $I = 0$; molal scale; solubility of $CaSO_4$ in dilute H_2SO_4 over a range of ionic strengths	SOLY	M35
2.03	30			
2.14	40			
2.18	45			
2.27	50			
2.36	60			
3.15	125			
3.56	150			
3.90	175			
4.24	200			
4.58	225			
4.98	250			
5.34	275			
5.71	300			
6.06	325			
6.41	350			
		$pK_2 = 19.8858 \log T + 0.006473T - 56.889 - 2307.9/T$ (T in ^{o}K)		
~-3.1		pK_1; theoretical prediction		S48
~-8.3		pK of $H_3SO_4^+$; theoretical prediction		
-3.3 to -3.5		pK_1; mole fraction equilibrium constant; prediction		W41
-3.0		pK_1; prediction, based on structure		K84
-1.7		pK_1; prediction; mole fraction equilibrium constant for $H_2O + H_2SO_4 \rightleftharpoons HSO_4^- + H_3O^+$		D44
-3.0		pK_1; prediction (Pauling's method)		G37
-2.0		pK_1; prediction		B5
-3.59		pK_1; from water activity data		W42
		For pK_1 values of H_2SO_4 from 400-800°C and densities from 0.40 to 0.85 $g.cm^{-3}$ (Pressures to 4700 bars), see Q7		
		For effect of pressure on pK_2, see H90		
-8.3		pK of $H_3SO_4^+$ (not hydrated)	RAMAN	H77

Name, Formula and pK value	T(^{o}C)	Remarks	Methods	Reference
-4.95	25	p$\underline{K}$ for 1:1 H_2SO_4:H_2O species		
		For self-dissociation constants of H_2SO_4 at 10, 25, 40°, see B17, G39, K36, S43		
		At 10.36^{o}, autoprotolysis constant of H_2SO_4 = 1.7 × 10^{-4}, for $2H_2SO_4 \rightleftharpoons H_3SO_4^+ + HSO_4^-$		G36
		At 10.36^{o}, ionic self-dehydration constant of H_2SO_4 = 7 × 10^{-5}, for $2H_2SO_4 \rightleftharpoons H_3O^+ + HS_2O_7^-$		
		For computed p$\underline{K}$ values for the reactions $H_2SO_4 + nH_2O \rightleftharpoons H^+.nH_2O + HSO_4^-$, and $HSO_4^- + nH_2O \rightleftharpoons H^+.nH_2O + SO_4^{2-}$, see R26		
		Other measurements of p$\underline{K}_2$ values: A12, A18, A52, B2, B50, B108, C15, C16, C24, C52, D28, D55, E5, E10, E33, F16, F26, F29, F41, G22, G65, H18, I29, K25, K41, K61, L43, L65, M23, M47, N45, P6, R7, R11, R20, R32, S33, S69, S80, S121, V8, V9, Z3		
		For Hammett acidity function of H_2SO_4, see B16, B103, D44, G27 and G29 (temperature range), H20, J18, P20, R57, S20		
		For H_R (J_o) acidity function of H_2SO_4, see A49, D42		
		For C_o acidity function of H_2SO_4, see D42		
		For $H_R{}^*$ acidity function of H_2SO_4, see Y9		
		For H_+ acidity function of H_2SO_4, see B104, V13		
		For H_A acidity function of H_2SO_4, see Y11, Y12		
		For $H_o{}'''$ acidity function of H_2SO_4, see A50		
		For $H_o{}'$ acidity function of H_2SO_4, see J18		
		For $H_R{}'$ acidity function of H_2SO_4, see B101		
		For H_- acidity function of H_2SO_4, see B100		
		For H_1 acidity function of H_2SO_4, see H73		
		For H_o acidity function of alcoholic H_2SO_4, see J1		

		For H_+ acidity function of alcoholic H_2SO_4, see I15 For R_o acidity function of H_2SO_4/H_2O, see B27 For H_o values of H_2SO_4/H_2O solutions, see V16 For temperature variation of H_o acidity function, see J11 For Hammett acidity functions of H_2SO_4/SO_3, H_2SO_4/HSO_3F, H_2SO_4/HSO_3Cl and $H_2SO_4/HB(HSO_4)_4$, see G88		
217. Sulphurous acid, H_2SO_3				
1.63	0	For I = 0; recalculation of data given in references C1 and M95	C1,R1b	J14
1.74	10			
1.81	18			
1.89	25			
1.98	35			
2.12	50			
1.51	0			D25
1.58	5			
1.64	10			
1.70	15			
1.77	20			
1.82	25			
1.76	25			R59
2.34	70			
2.62	100			
3.1	130			
3.5	150			
1.86	25	Apparent pK value; 1 atmosphere	C1	E19
1.69		500 atmospheres		
1.51		1000 atmospheres		
1.34		1500 atmospheres		
1.20		2000 atmospheres		

Name, Formula and pK value		T(^{o}C)	Remarks	Methods	Reference
1.764	7.205	25	$\underline{I}$ = 0.02 to 0.13 (for $p\underline{K}_1$), 0.014 to 0.12 (for $p\underline{K}_2$); extrapolated to $\underline{I}$ = 0	Elcg	T5
1.90	7.205	25		E3bg	Y21
1.86		25		C1	H100
1.86				O	
1.74		10	$p\underline{K}$ for $(SO_2 + H_2O)_{aq} \rightleftharpoons HSO_3^- + H^+$		A46
1.89		25			
2.00		35			
			$\underline{K}$ for $(SO_2.H_2O)_{aq} \rightleftharpoons (H_2SO_3)_{aq}$ = 0.47, 0.33 and 0.27 at 10°, 25° and 35° respectively.		
6.80		15	$p\underline{K}_2$; $\underline{I} \rightarrow 0$	E3bg	K94
7.00		20			
7.18		25			
7.26		30			
7.34		35			
7.41		40			
7.47		50			
7.50		60			
7.52		70			
7.54		80			
7.56		95			
			$p\underline{K}_1 = -1398/\underline{T} + 6.557$ ($\underline{T}$ in ^{o}K) $p\underline{K}_2$ = 7.22 For $p\underline{K}_2$ values in aqueous methanol solutions see D47 For values of $p\underline{K}_1$ and $p\underline{K}_2$ in D_2O/H_2O mixtures, see S10	LIT	E8
7.17		10	For $\underline{I}$ = 0; c <0.05; extrapolation by Debye-Huckel equation	E3bg	A45
7.30		25			
7.45		50			
6.96				E3bg	R53

			In aqueous solutions, bisulphite ions are in equilibrium with pyrosulphite ion; at 25°, $\underline{K} = 7 \times 10^{-2}$ for $[S_2O_5{}^{2-}]/[HSO_3{}^{-}]^2$		G52
			Other measurements: B118, B121, B122, C1, C66, D48, F8, F18, F50, K50, L48, M14, M95, S10, S33, S58, S69		
218. Tantalic acid					
9.6		18-20	Acidic $p\underline{K}$	SOLY	B1
13			Basic $p\underline{K}$		
219. Telluric acid, H_6TeO_6					
8.03	11.45	5	c = 0.005M H_6TeO_6; extrapolated to $\underline{I} = 0$ using simple Debye-Hückel relation	E3bg	E24
7.70	10.95	25			
7.59	10.80	35			
7.28	10.27	61			
			Thermodynamic quantities are derived from the results		
7.68	11.19	18	Titration of 0.04M acid	E3bg	B122
7.70	11.04	25	For $\underline{I} = 0$; at high concentrations, polytellurates are formed	E3bg	E1
	14.25	25	$p\underline{K}_3$	O7	
8.19		0	For $\underline{I} = 0$	E3b,quin	A37
7.98		10			
7.61		25			
7.43		35			
7.12		45			
			$p\underline{K}_1 = 8.180-2.36 \times 10^{-2}\ t$ (t in ^{o}C)		
7.60		25	c = 0.01M; polytellurates are formed when c is greater than 0.1	E3ag	J2
8.00		12	c = 0.06M H_6TeO_6	C1	F38
7.81		22			
7.63		32			

Name, Formula and pK value		T(°C)	Remarks	Methods	Reference
7.48		42			
7.37		50			
			For p$\underline{K}$ values of telluric acid in organic solvents, see K19		
7.47	9.34	22	successive p$\underline{K}$ values	E,C	B74
7.33		25	-log $\underline{K}$ for $H_6TeO_6 + H_2O \rightleftharpoons Te(OH)_7^- + H^+$; $\underline{I}$ = 1.5(KCl)		
6.38			-log $\underline{K}$ for $2H_6TeO_6 + H_2O \rightleftharpoons Te_2(OH)_{13}^- + H^+$		
13.44			-log $\underline{K}$ for $2H_6TeO_6 + 2H_2O \rightleftharpoons Te_2(OH)_{14}^{2-} + 2H^+$		
17.74			-log $\underline{K}$ for $H_6TeO_6 + 2H_2O \rightleftharpoons Te(OH)_8^{2-} + 2H^+$		
22.93			-log $\underline{K}$ for $2H_6TeO_6 + 3H_2O \rightleftharpoons Te_2(OH)_{15}^{2-} + 3H^+$		
7.32		25	p$\underline{K}$ for $H_2L \rightleftharpoons HL^- + H^+$	E3bg	B113
6.25			-log $\underline{K}$ for $2H_2L \rightleftharpoons H^+ + H_3L_2^-$		
13.23			-log $\underline{K}$ for $2H_2L \rightleftharpoons 2H^+ + H_2L_2^{2-}$		
7.61		25	For $\underline{I}$ = 0	E3b,R2a	A34
	>15		p$\underline{K}_3$		S98
			Other measurements: B83, B113, K2, L56, R35		
220. **Tellurium penta fluoride hydroxide**, $HOTeF_5$					
8.8			p$\underline{K}$ in glacial acetic acid	O	E29
221. (Aquo) **Tellurium(IV) ion**, Te^{4+}					
11.96		28	log $\underline{K}$ for *i* = 1, where for $Te^{4+} + i\ OH^- \rightleftharpoons Te(OH)_i^{(4-i)+}$	O8	N26
23.56			log $\underline{K}$ for *i* = 2		
34.83			log $\underline{K}$ for *i* = 3		
45.85			log $\underline{K}$ for *i* = 4		
222. **Tellurium(IV) isopolyacids**					
4.54	5.51	25	Stepwise p$\underline{K}$ values for ionization of $H_2Te_2O_5$		B93

4.68	7.60	25	Stepwise pK values for ionization of $H_2Te_4O_9$		

223. Tellurous acid, H_2TeO_3

2.46	3.16	25	concentration constants, $I = 0.1$ pK_1 for protonation of H_2TeO_3 pK_2 for mono-anion formation	DISTRIB	S58
3.5	5.4-5.8	25	pK_1 for protonation of H_2TeO_3; pK_2 for mono-anion formation	SOLY	I18
2			For protonation of H_2TeO_3; from thermodynamic data		L12
2.7	7.7	25	From hydrolysis of salts	O3	B83
2.8		20	$-\log K$ for $H_3TeO_3^+ \rightleftharpoons H^+ + H_2TeO_3$; $I \rightarrow 0$; $c < 10^{-3}M$ in TeO_3^{2-}	E3bg	M45
6.08		20	$-\log K$ for $H_2TeO_3 \rightleftharpoons H^+ + HTeO_3^-$		
9.96		20	$-\log K$ for $HTeO_3^- \rightleftharpoons H^+ + TeO_3^{2-}$		
6.15	8.40		successive pK values	E,SOLY	G8
6.92	9.43	20	$I = 0$	E3bg	N25
28.20		20	$\log K$ for $Te^{4+} + 2OH^- \rightleftharpoons Te(OH)_2^{2+}$, $I = 0.5$	E3bg	N3
14.0			$\log K$ for $Te(OH)_2^{2+} + OH^- \rightleftharpoons Te(OH)_3^+$		
11.23			$\log K$ for $Te(OH)_3^+ + OH^- \rightleftharpoons Te(OH)_4$		
			Other measurements: B95		

224. (Aquo) Terbium(III) ion, Tb^{3+}

8.16		25	pK_a for hydrolysis of Tb^{3+}; titration of 0.004-0.009 M $Tb(ClO_4)_3$ with 0.02M $Ba(OH)_2$; $I = 0.3$ ($NaClO_4$)	E3b	F49

225. Tetrachlorocobaltic acid, H_2CoCl_4

3.85		25	extraction into tributyl phosphate	DISTRIB	B42
3.83		25	extraction into dibutyl butyl phosphate		

Name, Formula and pK value		T(°C)	Remarks	Methods	Reference
226. Tetracyanohydroxo-oxomolybdic acid, $H_3MoO(OH)(CN)_4$					
8.81		25	pK_3; extrapolated to $I = 0$	E3bg	B33
8.86		30			
8.90		35			
8.97		40			
9.04		45			
9.13		50			
8.74		16			L49
			$pK_3 = 1730.72/T - 6.19 + 0.0308\,T$ (T in °K)		B33
227. Tetracyanonickelic acid, $H_2Ni(CN)_4$					
4.69	6.59	25	$I = 0.01–0.5$, extrapolated to $I = 0$		K0
4.74	6.69	40			
4.78	6.78	50			
4.48	5.40	25	$I = 0.3$		K48a
228. Tetrafluoroboric acid, HBF_4					
-0.44				ISOPIESTIC	S118
0.48			pK by extrapolation of apparent dissociation constant against dielectric constant of solvent		
2.77		25	pK, no details		S117
229. Tetra(hydrogen sulphato)arsenious acid, $HAs(HSO_4)_4$					
			For pK in H_2SO_4, see B14		
230. Tetra(hydrogen sulphato)boric acid, $HB(HSO_4)_4$					
			For pK in H_2SO_4, see B14		

231. Tetrametaphosphoric acid, $H_4P_4O_{12}$							
2.78				25	pK_4; for I = 0; from results at I ~0.01 (K^+ salt), using Debye-Hückel equation	E3bg	B53
2.74				25	pK_4; for I = 0; using assumed value for mobility of $HP_4O_{12}^{3-}$ ion	C2	D23
2.60	6.4	8.22	11.4		No details		K10
232. Tetramolybdic acid, $H_2Mo_4O_{13}$							
1.4	1.5				I = 1($NaClO_4$)	DISTRIB	C22
233. Tetraperoxychromic acid, H_3CrO_8							
7.16				30	pK_3; I = 3($NaClO_4$)	KIN	Q1
7.40				40			
7.60				50			
234. Tetrapolyphosphoric acid, $H_4P_4O_{13}$							
1.99	2.64	6.62	8.2	25	pK_1, pK_2, pK_3, pK_4, I = 0.034	E3bg	M24
1.91	2.56	6.75	8.50	50	I = 0.034		
1.79	2.41	6.65	8.51	60	I = 0.11		
1.69	2.33	6.72	8.53	65	I = 0.21		
1.82	2.39	6.92	8.90	70	I = 0.034		
1.36	2.23	6.63	8.34	25	pK_3, pK_4, pK_5, pK_6; "practical' constants; I = 1(NMe_4NO_3); NMe_4^+ salt; extrapolated against $(concentration)^{1/2}$; complex formation occurs with Na^+, K^+, or guanidinium ion	E3bg	W12
		7.38	9.11	25	pK_5, pK_6; for I = 0		
235. Tetrathiophosphoric acid, H_3PS_4							
1.5	3.5	6.6		20	I = 0	E3bg	P23

Name, Formula and pK value		T(^{o}C)	Remarks	Methods	Reference
236. (Aquo) Thallium(I) ion, Tl^{+}					
0.81		0	pK_b; I = 0.005 to 0.09; extrapolation to I = 0 using	SOLY	B38
0.82		25	Davies' equation; solubility of $TlIO_3$ in KOH solutions		
0.85		40			
			Thermodynamic quantities are derived from the results.		
0.85		25	pK_b; I = 0.02 to 0.05; extrapolation to I = 0 using Davies' equation	KIN	B39
0.42		25	pK_b; I = 0.08 to 0.25	CAT,KIN	B40
0.48		25	pK_b; molal scale; c = 0.0009-0.009	C2	L40c
0.69		25	pK_b; I = 0.5-5.0, extrapolated to I = 0	O	K104
237. (Aquo) Thallium(III) ion, Tl^{3+}					
1.07		25	pK for hydrolysis of Tl^{3+}; I = 1.5($NaClO_4$)	O6	R31
1.16		25	I = 3($NaClO_4$)		
1.01		40	I = 1.5($NaClO_4$)		
1.10		40	I = 3($NaClO_4$)		
-0.5		25	pK for hydrolysis of Tl^{3+}, assuming $TlOH^{2+}$, but not Tl^{3+},	KIN	H28
-0.7		32.2	can exchange with Tl^{+}; I = 6($HClO_4$ + $NaClO_4$)		
-0.8		41.8			
-0.8		25	pK for hydrolysis of Tl^{3+}, assuming $TlOH^{2+}$, but not	KIN	J9
-1.0		35	Tl^{3+}, can exchange with Tl^{+}; I = 3($HClO_4$ + $NaClO_4$);		
-1.1		45	data of R.J. Prestwood and A.C. Wahl, J. Am. Chem. Soc., 71, 3137 (1949).		
1.14	1.49	25	pK values for successive hydrolysis of Tl^{3+}; I = 3($NaClO_4$); from potentials of Tl^{+}/Tl^{3+} electrode		B57
1.10			pK for $Tl^{3+} + H_2O \rightleftharpoons TlOH^{2+} + H^{+}$		K107
2.60			$-\log K$ for $Tl^{3+} + 2H_2O \rightleftharpoons Tl(OH)_2^{+} + 2H^{+}$		
			for hydrolysis constants in dimethylsulphoxide/water, dioxian/water and acetonitrile/water, see K102, K107.		

				Other measurements: K103		
237. Thiocyanic acid, HSCN						
-1.1			20	HCl + NH_4SCN; I = 2.4–5.1	RAMAN	C54
~-2				By interpolation from pK values in water for HBr, HCl, HNO_3, and HF, and in ethanol for HBr, HCl, HNO_3, HCNS and HF		M96
-1.85			25	From solvent (CCl_4) extraction of a series of $NaClO_4$, $HClO_4$ mixtures at constant I; extrapolated against I.	DISTRIB	
~-1.4			25	From H_- dependence of Fe^{3+} – CNS^- reaction and hydrolysis of NONCS	O6	
-0.7				I = 3; 2.3 M in $HClO_4$	KIN	T25
				Other measurements: B98, S126		
238. Thiophosphoric acid (Phosphorothioic acid), H_3PO_3S						
	5.83		25	pK_2, I = 0.01-1.9 (KCl), extrapolated to I = 0	E3bg	M21
<2	5.75	10.4	23		E3bg	N31
	6.0	10.8			O1	
	5.91	10.51			E3bg	F15
1.265	5.280	10.29	10	I = 0.5, extrapolated to I = 0	E3bg	P22
1.517	5.390	10.19	16			
1.788	5.427	10.08	25			
-	5.428	10.00	32.2			
239. Thiosulphuric acid, $H_2S_2O_3$						
0.6	1.74		25	I = 0.09 to 0.91, extrapolated to I = 0 using Davies' equation	E3bg	P3
	1.46		25	Concentration constant; I = 0.016 to 0.07	E3ag	D38
	1.56		25		E3bg	Y25

Name, Formula and pK value				T(^{o}C)	Remarks	Methods	Reference
					Other measurements: J3, K56		
240. (Aquo) Thorium(IV) ion, Th^{4+}							
3.89	4.20			25	Successive p$\underline{K}$ values for hydrolysis of Th^{4+}, assuming only mononuclear species: $\underline{I}$ = 0.05 to 0.5($NaClO_4$); c = 0.0001-0.01 M $Th(NO_3)_4$; extrapolated to $\underline{I}$ = 0	E3ag	P11
4.3	3.4			25	Successive p$\underline{K}$ values for hydrolysis of Th^{4+} at very slight degrees of hydrolysis; $\underline{I}$ = 1($NaClO_4$); also $-\log \underline{K} = 4.7$ for $2Th^{4+} + 2H_2O \rightleftharpoons Th_2(OH)_2^{6+} + 2H^+$	E3ag	K83
4.32	4.16			0	Successive p$\underline{K}$ values for hydrolysis of Th^{4+}; $\underline{I}$ = 1($NaClO_4$); also $-\log \underline{K} = 5.60$ for $2Th^{4+} + 2H_2O \rightleftharpoons Th_2(OH)_2^{6+} + 2H^+$; $-\log \underline{K} = 22.79$ for $Th_4(OH)_6^{10+}$; $-\log \underline{K} = 43.84$ for $Th_6(OH)_{15}^{9+}$	E2a,quin	B4
4.12	3.69			25	$-\log \underline{K} = 4.61$ for $Th_2(OH)_2^{6+}$; $-\log \underline{K} = 19.01$ for $Th_4(OH)_6^{10+}$; $-\log \underline{K} = 36.76$ for $Th_6(OH)_{15}^{9+}$		
2.29	2.21			95	$-\log \underline{K} = 2.55$ for $Th_2(OH)_2^{6+}$; $-\log \underline{K} = 10.49$ for $Th_4(OH)_6^{10+}$; $-\log \underline{K} = 20.63$ for $Th_6(OH)_{15}^{9+}$	E2ag	
11.64	10.80	10.62	10.45	25	$\underline{I}$ = 0.5($NaClO_4$), radio-tracer Concentration of Th(IV); successive p$\underline{K}_b$ values.	ION	B43
9.40	9.35	8.99	8.08		$\underline{I}$ = 0.1($NaClO_4$); successive p$\underline{K}_b$ values	SOLY	N4
					At 25° and $\underline{I}$ = 3(NaCl), c = 0.0001-0.1 M in Th^{4+}, $-\log \underline{K} = \sim 9.1$ for $Th^{4+} + 2H_2O \rightleftharpoons Th(OH)_2^{2+} + 2H^+$; $ThOH^{3+}$ is negligible; Th_2OH^{7+}, $Th_2(OH)_2^{6+}$, $Th_3(OH)_2^{10+}$ and higher complexes are important; constants are given.		H68
					At high thorium concentrations (0.5 M) formation of $Th_2(OH)_2^{6+}$ and Th_2OH^{7+} is important; constants are given		H66
3.15	6.56			25	Successive p$\underline{K}$ values for hydrolysis of Th^{4+}; $\underline{I}$ = 0.05	E3bg	U2

		($NaCl, NaClO_4$); c = 0.005 M Th^{4+}		
3.71	25	pK for hydrolysis of Th^{4+}; I = 1($NaClO_4$)		H69
4.44	25	−log K for $2Th^{4+} + 2H_2O \rightleftharpoons Th_2(OH)_2^{6+} + 2H^+$; I = 1($NaClO_4$)		
5.14	25	−log K for $2Th^{4+} + 2H_2O \rightleftharpoons Th_2(OH)_2^{6+} + 2H^+$; I = 3($LiNO_3$, KNO_3 or $Mg(NO_3)_2$)	E3bg	M75
		Complexes also formed include $Th_3(OH)_5$, $Th_2(OH)_3$ and $Th_6(OH)_{15}$		
4.60	25	pK for hydrolysis of Th^{4+}; I = 3(NaCl); c = 0.0001–0.1M Th^{4+}	E3bh	H69
4.83	25	−log K for $2Th^{4+} + 2H_2O \rightleftharpoons Th_2(OH)^{6+} + 2H^+$		
		Hexanuclear complexes are also formed.		
		Other measurements: D4, K12, L20, S60.		

241. (Aquo) Thulium(III) ion, Tm^{3+}

	25	pK_a for hydrolysis of Tm^{3+}; titration of 0.004-0.009 M $Tm(ClO_4)_3$ with 0.02 M $Ba(OH)_2$; I = 0.3($NaClO_4$)	E3b	F49

242. (Aquo) Tin(II) ion, Sn^{2+}

1.70	25	pK for hydrolysis of Sn^{2+}; I = 0.14 to 0.5, extrapolated against $I^{1/2}$; c = 0.004-0.12 M Sn^{2+}	E3ah	G56
11.93	25	pK_b for $SnOH^+ \rightleftharpoons Sn^{2+} + OH^+$	SOLY	G9a
1.82	0	pK for hydrolysis of Sn^{2+}; I = 3($NaClO_4$);		V3
1.70	25	Sn-Hg electrode potentials		
1.64	35			
1.60	45			
3.92	25	pK for hydrolysis of Sn^{2+}; I = 3($NaClO_4$); also −log K = 4.45 for $2Sn^{2+} + 2H_2O \rightleftharpoons Sn_2(OH)_2^{2+} + 2H^+$, −log K = 6.77 for $3Sn^{2+} + 4H_2O \rightleftharpoons Sn_3(OH)_4^{2+} + 4H^+$,	E3bg	T11
3.2	25	pK for hydrolysis of Sn^{2+}; I = 2($NaNO_3$)	E3bg	D26

Name, Formula and pK value	T(^{o}C)	Remarks	Methods	Reference
3.70	25	p$\underline{K}$ for hydrolysis of Sn^{2+}; $\underline{I}$ = 3($NaClO_4$); c = 0.02–2.3 mM Sn^{2+}	E3bg, Sn(Hg)	G46
6.81	25	–log $\underline{K}$ for $3Sn^{2+} + 4H_2O \rightleftharpoons Sn_3(OH)_4{}^{2+} + 4H^+$; $\underline{I}$ = 3($NaClO_4$)		
24.58	25	–log $\underline{K}$ for $Sn^{2+} + 3OH^- \rightleftharpoons Sn(OH)_3{}^-$; $\underline{I}$ = 3($NaClO_4$) Other measurements: P64, S90	E3b, Sn(Hg)	M32
243. (Aquo) Tin(IV) ion				
14.15	25	–log $\underline{K}$ for $Sn^{4+} + iOH^- \rightleftharpoons Sn(OH)_i{}^{(4-i)+}$, i = 1; $\underline{I}$ = 1	O2	N19
28.68	25	i = 2		
42.35	25	i = 3		
55.13	25	i = 4		
244. (Aquo) Titanium(III) ion, Ti^{3+}				
1.41	15	p$\underline{K}$ for $Ti^{3+} + H_2O \rightleftharpoons TiOH^{2+} + H^+$; $\underline{I}$ = 0.25 to 1.5(KBr);	E3bg	P26
1.29	25	extrapolated to $\underline{I}$ = 0 using modified Debye-Hückel equation		
1.36	35	At pK values above 3.5 polynuclear species are rapidly formed.		
2.55	25	p$\underline{K}$ for hydrolysis of Ti^{3+} to $TiOH^{2+} + H^+$; $\underline{I}$ = 3(KBr)	E3bg	P15
3.30	25	–log $\underline{K}$ for $2Ti^{3+} + 2H_2O \rightleftharpoons Ti_2(OH)_2{}^{4+} + 2H^+$; $\underline{I}$ = 3(KBr)		
2.55		p$\underline{K}$ of Ti^{3+}; $\underline{I}$ = 3($NaClO_4$) Other measurements: K89	REDOX	B67
245. (Aquo) Titanium(IV) ion, Ti^{4+}				
1.8 2.4 2.1	25	Successive p$\underline{K}$ values for hydrolysis of $TiOH^{3+}$; $\underline{I}$ = 0.1($NaClO_4$); c = 10^{-4}–5 × 10^{-4} M in Ti(IV)	DISTRIB	L36
1.3	18	p$\underline{K}$ for $TiO^{2+} \rightleftharpoons TiO(OH)^+ + H^+$	ION	N1

0.3						$p\underline{K}$ for $TiO^{2+} \rightleftharpoons TiO(OH)^{+} + H^{+}$; in $HClO_4$ solutions; in dilute H_2SO_4 basic sulphates are also formed	ION	B52
1.68					25	$-\log \underline{K}$ for $8TiO^{2+} + 12H_2O \rightleftharpoons (TiO)_8(OH)_{12}^{4+} + 12H^{+}$; $\underline{I} = 2(NaCl,HCl)$; $c = 0.02=0.05M\ Ti^{4+}$ Other measurements: N5	E3ag	E13
246. Trichlorocobaltic acid, $HCoCl_3$								
2.98					25	using tributyl phosphate	DISTRIB	B42
2.57					25	using dibutyl butyl phosphate		
247. Trimetaphosphoric acid, $H_3P_3O_9$								
1.74					25	$p\underline{K}_3$; inversion of sucrose	KIN	I1
2.05					25	$p\underline{K}_3$; for $\underline{I} = 0$; using an assumed value for the mobility of $HP_3O_9^{2-}$ ion	C2	D23
1.64	2.07				25	$\underline{I} = 0.2\ (NH_4ClO_4)$	E3bg	E14
248. Triphophoric acid, $H_5P_3O_8$								
7.32					25	$p\underline{K}$ for $H_2P_3O_8^{3-} \rightleftharpoons HP_3O_8^{4-} + H^{+}$; $\underline{I} = 0.1$	E3bg	S88
	10.28				25	$p\underline{K}$ for $HP_3O_8^{4-} \rightleftharpoons P_3O_8^{5-} + H^{+}$		
249. Tripolyphosphoric acid, $H_5P_3O_{10}$								
0.51	1.20				25	For $\underline{I} = 0$	CALORIM	I14
~1	2.2	2.3	5.7	8.51	0	Concentration constants; $p\underline{K}_1$ and $p\underline{K}_2$ are very uncertain; NMe_4^{+} salts; $\underline{I} = 0.1\ (NMe_4Br)$; $f\pm$ assumed same as for HBr	E3bg	I12
~1	2.2	2.31	5.84	8.70	10			
~1	2	2.13	5.75	8.65	25			
~1	1.7	1.89	5.77	8.50	37			
~1	1.7	1.95	5.77	8.51	37	$\underline{I} = 0.2$		
~1	1.7	1.98	5.78	8.52	37	$\underline{I} = 0.3$		
~1	1.7	2.12	5.90	8.55	50	$\underline{I} = 0.1$		
~1	1.7	1.95	5.80	8.48	50	$\underline{I} = 0.2$		

Name, Formula and pK value					T(^{o}C)	Remarks	Methods	Reference
~1	1.7	2.62	5.84	8.48	50	I = 0.3		
~1	1.7	2.15	5.88	8.48	65	I = 0.1		
~1	1.7	2.10	5.80	8.39	65	I = 1.0		
~0.5	1.15	2.04	5.69	8.56	25	I = 1 (NMe_4Br); as for I12	E3bg	I13
	1.06	2.11	5.83	8.81	25	"Practical" constants: NMe_4^+ salt; I = 1.0 (NMe_4Cl)	E3bg	L7,W11
		2.15	6.00	8.73		I = 0.1 (NMe_4Cl)		
		2.30	6.50	9.24		For I = 0, by extrapolation against (concentration)$^{1/2}$		
	2.2	2.6	5.6	7.9	25	"Practical" constants: I = 0.1 ($NaNO_3$); $c = 10^{-3}$ M	E3bg	J8
		2.79	6.47	9.24	25	At I = 0, extrapolated from I ~0.01 (K^+ salt), using Debye-Hückel equation	E3bg	B53
			5.29	7.58	27.4	"Practical" constants; I = 0.75 ($NaNO_3$)	E3bg	Y8
			5.43	8.06	25	Concentration constants: I = 0.1 (KCl); c = 0.001 M	E3bg	E25
			5.43	7.87	20	Concentration constants; I = 0.1 (KCl); c = 0.001 M	E3bg	M37
9.54					25	pK_5; I = 0.005 to 0.04, extrapolated to I = 0	E3bh	W32
9.62					40			
0.75	1.19	2.20	5.92	8.88	25	I = 0.1 (NMe_4Br); values are also given for I = 0.4, 0.7, 1.0	E3bg	M84
	1.30	2.25	6.15	9.04	37			
0.74	1.10	2.15	6.05	8.95	50			
0.56	1.15	2.01	6.14	9.05	60			
0.70	1.02	2.0	6.26	9.13	70			
			5.79	8.81	25	I = 0.1 (NMe_4Cl)	E3bg	M13
			6.61	9.26	20	Corrected to I = 0	E3bg	P45
		~2.2	5.93	8.82	20	I = 0.1 (NMe_4NO_3)	E3bg	A31
						Other measurements: E7		
250. Triselenocarbonic acid, H_2CSe_3								
	1.16		7.70		0	Corrected to I = 0	C1	G16

		7.50	10			
		7.41	15			
		7.28	20			
		7.13	25			
251. Trithiocarbonic acid, H_2CS_3						
	2.81		4.5	c = 0.005-0.02 N; extrapolated against $\underline{I}^{1/2}$	C1	G17
	2.76		8			
	2.72		14			
	2.68		20			
		8.18	20	Hydrolysis of 0.005 and 0.01 M Na_2CS_3 and K_2CS_3 solutions	E3ag	
252. Tritium oxide, T_2O						
15.21			25	$p\underline{K}_w$ for $T_2O \rightleftharpoons T^+ + OT^-$	E3at	G50
253. Tungstic acid, H_2WO_4. See also Dodecatungstic acid						
~3.5	~4.6		20	$\underline{I}$ = 0.1($NaClO_4$); rapid-reaction technique Under ordinary conditions, WO_4^{2-} is in equilibrium with $HW_6O_{11}^{5-}$; log $\underline{K}$ = 65.5 at 25° for $(HW_6O_{21}^{5-})(H_2O)^3/(H^+)^7(WO_4^{2-})^6$ (ref. D66) or = 60.7 at 25° in 3 M $NaClO_4$ (ref. S15)	E3ag	S52
~2.3	~3.5		25	$p\underline{K}_1$, $p\underline{K}_2$; var.	KIN	Y17
2.20	3.70		25	$p\underline{K}_1$, $p\underline{K}_2$; $\underline{I}$ = 0.003 $p\underline{K}$ for $H^+ + H_2WO_4 \rightleftharpoons H_3WO_4^+$		
		6.84		$p\underline{K}_2$		C30
				log $\underline{K}$ for $HW_6O_{21}^{5-} + 2H^+ \rightleftharpoons H_3W_6O_{21}^{3-}$ = 9.72; $\underline{I}$ = 1 log $\underline{K}$ for $6WO_4^{2-} + 9H^+ \rightleftharpoons H_3W_6O_{21}^{3-} + 3H_2O$ = 71.1 Least-squares refined equilibrium constants (for 3 M	E3bg	T14

Name, Formula and pK value	T(°C)	Remarks	Methods	Reference
		LiCl at 50°) are		
		$7H^+ + 6WO_4^{2-} \rightleftharpoons 3H_2O + HW_6O_{21}^{5-}$, log K = 53.98		
		$14H^+ + 12WO_4^{2-} \rightleftharpoons 7H_2O + W_{12}O_{41}^{10-}$, log K = 110.03		
		$18H^+ + 12WO_4^{2-} \rightleftharpoons 9H_2O + W_{12}O_{39}^{6-}$, log K = 132.51		A57
		For thermochemical studies, see A48		
		Other measurements: C31		
254. (Aquo) Uranium(IV) ion, U^{4+}				
1.90	10	pK for hydrolysis of U^{4+} to UOH^{3+}; I = 0.5($NaClO_4$, $HClO_4$)	O6	K86
1.00	43	$c = 7 \times 10^{-4}$ M U^{4+}		
~1.12	10	Extrapolation to I = 0, using modified Debye-Hückel equation		
0.68	25	at I = 0		
~0.18	43	at I = 0		
0.68	25	pK for hydrolysis of U^{4+}; I = 0.02 to 2.0; extrapolated to I = 0 by fitting to extended Debye-Hückel equation	O6	K85
1.38	15.2	pK for hydrolysis of U^{4+}; I = 0.19($HClO_4$)	O6	B51
1.12	24.7			
1.68	25	I = 2($NaClO_4$)	O5	S120
1.74	25	I = 2($NaClO_4$); D_2O solution		
2.00	25	I = 3($NaClO_4$); similar value from redox potentials; poly-nuclear complexes are also formed	E3bg,h	H63
12.5		pK_b for hydrolysis of U^{4+}; from spin-lattice relaxation time *vs* pH in 0.03 M solutions	NMR	V10
		For pK values in ethanol-water mixtures, see R24		
1.11	25	pK for hydrolysis of U^{4+}; I = 0		S94
1.28	20	pK for hydrolysis of U^{4+}	O	G44
1.57	20	pK for hydrolysis of U^{4+}; I = 1($LiClO_4$)	O	M12
		pK = 2178.44/T −6.89, where T is in °K for 25-150°C,		

				$\underline{I}$ = 0.25-2.0, for $U^{4+} + H_2O \rightleftharpoons UOH^{3+} + H^+$ Other measurements: D11, L15	O	N39
255. (Aquo) Uranyl ion, UO_2^{2+}						
5.82			25	$p\underline{K}$ for hydrolysis of UO_2^{2+}; from hydrolysis of pure salts; corrected for formation of $(UO_2)_2(OH)_2^{2+}$; $\underline{I}$ = 0.035($NaClO_4$)	E3ag	H49
5.05			16	$p\underline{K}$ for hydrolysis of UO_2^{2+}; from hydrolysis of pure	E3ag	P59
4.59			27	salts; c = 0.001-0.015 M uranyl nitrate; at higher concentrations hydroxonitrato complexes are formed		
5.70			25	$p\underline{K}$ for hydrolysis of UO_2^{2+}; $\underline{I}$ = 0.5(KNO_3); constants	E3bg	B3
4.19			94.4	are also given for $(UO_2)_2(OH)_2^{2+}$ and $(UO_2)_3(OH)_5^{+}$, which are major species in 0.01 M solutions		
6.10			25	$p\underline{K}$ for hydrolysis of UO_2^{2+}; allowance was made for formation of $(UO_2)_2(OH)_2^{2+}$	E3bg	G81
6.34				$p\underline{K}_b$ for $UO_2(OH)_2 + OH^- \rightleftharpoons UO_2OH^- + H_2O$; corrected for formation of $(UO_2)_2(OH)_2^{2+}$, $(UO_2)_4(OH)_6^{2+}$ and $U_2O_7^{2-}$	E3b	I17
9.2	8.0	8.3	20	Successive $p\underline{K}_b$ values for formation of UO_2OH^+, $UO_2(OH)_2$ and $UO_2(OH)_3^-$; $\underline{I}$ = 0.1($NaClO_4$)	DISTRIB	S109
5.98			25	$p\underline{K}$ for hydrolysis of UO_2^{2+}; determined by dissociation field effect relaxation method	KIN	C41
				The major species in hydrolysed UO_2^{2+} solutions, 1 M in KCl or $NaClO_4$, are $(UO_2)_2(OH)_2^{2+}$, $(UO_2)_3(OH)_4^{2+}$ and $(UO_2)_3(OH)^{5+}$; constants are given Results in 3 M NaCl, $Mg(ClO_4)_2$, $Ca(ClO_4)_2$, and $NaClO_4$, 1 M KNO_3, and 1.5 M Na_2SO_4 are interpreted in terms of five or more polynuclear complexes for which equilibrium constants are given.		R55
6.52			25	$-\log \beta_{22}$ for $2UO_2^{2+} + 2H_2O \rightleftharpoons (UO_2)_2(OH)_2^{2+} + 2H^+$;	E,g	S27

Name, Formula and pK value	T(^{o}C)	Remarks	Methods	Reference
		$\underline{I} = 5(Mg(NO_3)_2)$		
17.76		$-\log \beta_{35}$ for $3UO_2^{2+} + 5H_2O \rightleftharpoons (UO_2)_3(OH)_5^{+} + 5H^{+}$		
5.53		$-\log \beta_{11}$		
6.03	25	$-\log \beta_{22}$; $\underline{I} = 0.5(NaClO_4)$	E,g	L3
16.78		$-\log \beta_{35}$		
3.81		$-\log \beta_{21}$		
13.17		$-\log \beta_{34}$		
18.91		$-\log \beta_{46}$		
5.50	25	$-\log \beta_{11}$; $\underline{I} = 0.1(KNO_3)$	E3bg	S131
5.89		$-\log \beta_{22}$		
12.31		$-\log \beta_{34}$		
16.46		$-\log \beta_{35}$		
22.76		$-\log \beta_{47}$		
6.80	25	$-\log \beta_{22}$; in D_2O; $\underline{I} = 3(NaClO_4)$; 1–10 mM in UO_2^{2+}	E3bg	M16
18.63		$-\log \beta_{35}$		
14.10		$-\log \beta_{34}$		
		(In H_2O the constants are 6.04, 16.53, 13.21, respectively, and $-\log \underline{K} = 6.10$)		
		Other measurements: A17, D10, G20, G77, G78, H45, H53, K66, M50, N38, P56, R56, R58, V20		
256. <u>Vanadic acid,</u> $H_3VO_4(HVO_3)$.		See also Decavanadic acid		
		At pH values less than 1, pentavalent vanadium exists as the cation VO_2^{+}. In less acid solutions it undergoes extensive aggregation to polynuclear species, including decavanadates.		
3.78	25	$\underline{I} = 0.5(NaClO_4)$; using tracer concentrations of ^{48}V; $-\log \underline{K} = 3.20$ for $VO_2^{+} + H_2O \rightleftharpoons HVO_3 + H^{+}$	DISTRIB	D74

3.4	8.23		20	Rapid-reaction studies; vanadate solutions initially at pH 9-10; final vanadate concentration 2-4 × 10^{-4} M; apparent p$\underline{K}$; $\underline{I}$ = 0.1($NaClO_4$)	O5	S51
	8.31		20	ditto; $\underline{I}$ = 0.1(NMe_4Cl)		
	7.8		25	$\underline{I}$ = 0.5(NaCl); log $\underline{K}$ = 7.6 for $3H_2VO_4^- \rightleftharpoons V_3O_9^{3-} + 3H_2O$;		
				–log $\underline{K}$ = 5.0 for $2H_2VO_4^- \rightleftharpoons HV_2O_7^{3-} + H^+ + H_2O$		
		13.2	25	$\underline{I}$ = 3($NaClO_4$); $\underline{K}$~50 for $2HVO_4^{2-} \rightleftharpoons V_2O_7^{4-} + H_2O$		
	7.72	13.0	25	$\underline{I}$ = 0.5(NaCl); log $\underline{K}$ = -3.18 for $2HVO_4^{2-} \rightleftharpoons HV_2O_7^{3-} + OH^-$; log $\underline{K}$ = -10.42 for $3HVO_4^{2-} \rightleftharpoons V_3O_9^{3-} + 3OH^-$	E3bg	I8
	7.83		25	$\underline{I}$ = 0.5; recalculation of earlier data; constants are given for several condensed species		B114,I8
	8.44			Also log $\underline{K}$ = -5.22 for $V_3O_9^{3-} \rightleftharpoons 3VO_3^-$;	O5	S29
				–log $\underline{K}$ = 3.5 to 4.0 for $VO_2^+ \rightleftharpoons H^+ + HVO_3$;		
				–log $\underline{K}$ = 4.3 to 4.8 for $HVO_3 \rightleftharpoons H^+ + VO_3^-$		
	8.95		25	Dilute solutions	O5	B5
		14.4	25		O7	
	9.62	12.72	32.4	$\underline{I}$ = 9; in saturated Na_2SO_4 solutions	CRYOSC	S31
		11.13	20	$\underline{I}$ = 0.2; in acid solutions V exists as VO_2^+ and VO^{3+}, polymerizing in neutral solutions to decavanadates. In alkaline solutions, mononuclear species predominate if $c \leqslant 10^{-3}$ M;	E3ag	S12
				log $\underline{K}$ = 25.05 for $2VO_4^{3-} + 2H^+ \rightleftharpoons V_2O_7^{4-} + H_2O$;		
				log $\underline{K}$ = -2.79 for $2HVO_4^{2-} \rightleftharpoons V_2O_7^{4-} + H_2O$;		
				log $\underline{K}$ = -10.99 for $\frac{1}{4}V_4O_{12}^{4-} + H_2O \rightleftharpoons HVO_4^{2-} + H^+$		
		12.99	25	Concentration constant; $\underline{I}$ = 3($NaClO_4$);	O5	N35
				$\underline{K}$ = 48 for $2HVO_4^{2-} \rightleftharpoons V_2O_7^{4-} + H_2O$		
		13.15	33	$\underline{I}$ = 9; saturated Na_2SO_4 solutions	CRYOSC	S100
				$\underline{K}$ = 2.0 × 10^{-4} for $H_3V_2O_7^- + 3H^+ \rightleftharpoons 2VO_2^+ + H_2O$; important between pH2 and pH4.		T15

Name, Formula and pK value			(T^oC)	Remarks	Methods	Reference
				$\underline{K} = 1.2 \times 10^3$ for $H_3V_2O_7^- + 3H^+ \rightleftharpoons 2VO_2^+ + H_2O$		D63
				$\underline{K} = 2.8 \times 10^{-4}$ for $H_3V_2O_7^- + H^+ \rightleftharpoons 2HVO_3 + H_2O$, at		Y15
				$\underline{I} = 0.006$ and 25°		
				$\underline{K} = 1.64 \times 10^3$ for $HVO_3 + H^+ \rightleftharpoons VO_2^+ + H_2O$		
				$\log \underline{K} = -30.48$ for $V_3O_9^{3-} \rightleftharpoons 3H^+ + 3HVO_4^{2-}$		S97
				Also, $\log \underline{K} - 11.09$ for $2HVO_4^{2-} + H^+ = HV_2O_7^{3-}$ +	O	B94
				H_2O; $\log \underline{K} = 33.51$ for $3HVO_4^{2-} + 3H^+ = V_3O_9^{3-} + 3H_2O$;		
				$-\log \underline{K} = 12.09$ for $HVO_4^{2-} \rightleftharpoons VO_4^{3-} + H^+$;		
		10.86	10	$p\underline{K}_3$; $\underline{I} = 3(NaClO_4)$, 0.1$\underline{M}$ TRIS buffer	O	R22
	7.62		25	$p\underline{K}_2$; $\underline{I} = 3(NaClO_4)$, 0.1$\underline{M}$ TRIS buffer		
	7.03		40			
				Other measurements: B111, D63, L19, Z5		
257. (Aquo) Vanadium(II) ion, V^{2+}						
	6.85		25	VSO_4; $p\underline{K}$ for $V^{2+} + H_2O \rightleftharpoons VOH^+ + H^+$	E,h	P52
	6.49		30			
	6.10		35			
258. (Aquo) Vanadium(III) ion, V^{3+}						
2.92	3.5		25	Successive $p\underline{K}$ values for hydrolysis of V^{3+}; calculated	E3ag	M53
				from data of G. Jones and W.A. Ray, J. Am. Chem. Soc.,		
				66, 1571 (1944);		
				$c = 0.0004$-0.04 M $V_2(SO_4)_3$		
2.4	3.85		22.5	Successive $p\underline{K}$ values for hydrolysis of V^{3+}; $\underline{I} = 1(NaCl)$;	E	P4
				also $-\log \underline{K} = 3.90$ for $2V^{3+} + 2H_2O \rightleftharpoons V_2(OH)_2^{4+} + 2H^+$		
2.7						F53
		12.98	20	$\log \underline{K}$ for $V^{3+} + OH^- \rightleftharpoons VOH^{2+}$, $\underline{I} = 0.1$	DISTRIB	S65
		25.52	20	$\log \underline{K}$ for $V^{3+} + 2OH \rightleftharpoons V(OH)_2^+$		
		37.62	20	$\log \underline{K}$ for $V^{3+} + 3OH^- \rightleftharpoons V(OH)_3$		

3.07	25	pK for $V^{3+} + H_2O \rightleftharpoons VOH^{2+} + H^+$; $I = 3(KCl)$	E,h	B112,D65
		also $-\log K = 3.93$ for $2V^{3+} + 2H_2O \rightleftharpoons V_2(OH)_2^{4+} + 2H^+$,		
		$-\log K = 8.0$ for $2V^{3+} + 3H_2O \rightleftharpoons V_2(OH)_3^{3+} + 3H^+$		
		Other measurements: B111		
259. (Aquo) Vanadyl ion, VO^{2+} (= $V(OH)_2^{2+}$)				
4.77	20	pK for $VO^{2+} + H_2O \rightleftharpoons VO{\cdot}OH^+ + H^+$; $c = 0.012$ M VO^{2+};	E3bg	D63
		no correction for dimerization		
5.36	25	pK for VO^{2+}; $c = 0.0001$-0.5 M $VOSO_4$; calculated	E3ag	M53
		from data of G. Jones and W.A. Ray, J. Am. Chem. Soc.,		
		66, 1571 (1944)		
7.36	25	pK for $VO^{2+} + H_2O \rightleftharpoons VO{\cdot}OH^+ + H^+$	OTHER	F39
		electron paramagnetic resonance		
6.82		$-\log K$ for $2VO^{2+} + 2H_2O \rightleftharpoons (VO{\cdot}OH)_2^{2+} + 2H^+$		
21.97		$\log K$ for $VO^{2+} + 2OH^- \rightleftharpoons VO(OH)_2$		
6.88	25	pK for VO^{2+}; $I = 3(NaClO_4)$; $\log K = 5.1$ for dimerization	E3bg	R41
		of $VO{\cdot}OH^+$ to $(VO)_2(OH)_2^{2+}$		
−0.5	Room	pK for $VOH^{3+} \rightleftharpoons VO^{2+} + H^+$; from changes in nuclear		R23
		relaxation times; in dilute H_2SO_4		
−0.6		in dilute HNO_3		
−0.9		in dilute $HClO_4$		
5.05	25	pK for VO^{2+}; $I = 0.1(LiClO_4)$	E3bg	K70
6.73		$-\log K$ for $2VO^{2+} + H_2O = (VO)_2(OH)^{2+} + 2H^+$		
7.64	4	$-\log K$ for $2VO^{2+} + H_2O = (VO)_2(OH)^{2+} + 2H^+$; $I = 0.3$		L66
7.22	16	$(NaClO_4)$		
6.95	25			
6.60	36			
6.30	46			
		Other measurements: B44, H57		

Name, Formula and pK value	T($^{\circ}$C)	Remarks	Methods	Reference
260. Vanadium(V) ion, VO_2^+				
1.83	20	p$\underline{K}$ for $VO_2^+ + H_2O \rightleftharpoons HVO_3 + H^+$; $\underline{I}$ = 0.1	DISTRIB	S65
2.38	20	$-\log \underline{K}$ for $VO_2^+ + 2H_2O \rightleftharpoons VO_2(OH)_2^- + 2H^+$		
7.63		$-\log \underline{K}$ for $10VO_2^+ + 8H_2O \rightleftharpoons H_2V_{10}O_{28}^{4-} + 14H^+$; $\underline{I}$ = 1(NaClO$_4$)	O1	B94
11.57		$-\log \underline{K}$ for $10VO_2^+ + 8H_2O \rightleftharpoons HV_{10}O_{28}^{5-} + 15H^+$		
17.40		$-\log \underline{K}$ for $10VO_2^+ + 8H_2O \rightleftharpoons V_{10}O_{28}^{6-} + 16H^+$		
261. Water, H_2O				
14.9435	0	$\underline{I}$ = 0 re-examination of data by Harned and co-workers;	Elch	H43
14.7338	5	molal scale		
14.5346	10			
14.3463	15			
14.1669	20			
13.9965	25			
13.8330	30			
13.6801	35			
13.5348	40			
13.3960	45			
13.2617	50			
13.1369	55			
13.0171	60			
14.535	10	Molar scale		
14.169	20			
14.000	25			
13.837	30			
13.542	40			
13.272	50			
16.279	10	Mole fraction scale		

15.911	20			
15.741	25			
15.577	30			
15.279	40			
15.006	50			
14.946	0	LiBr solutions; *I* = 0.01 to 3.0;	Elch	H37
14.735	5	E corrected using Debye-Hückel equation and		
14.535	10	extrapolated against *I* to *I* = 0		
14.346	15			
14.167	20			
13.997	25			
13.834	30			
13.680	35			
13.539	40			
14.945	0	NaCl solutions; as for H37	Elch	H41
14.535	10			
14.167	20			
13.997	25			
13.833	30			
13.536	40			
13.261	50			
13.015	60			
14.939	0	KCl solutions; *I* = 0.01 to 3.5 as for H37	Elch	H39
14.730	5			
14.533	10			
14.345	15			
14.167	20			
13.997	25			
13.832	30			
13.620	35			
13.535	40			

Name, Formula and pK value	T(^{O}C)	Remarks	Methods	Reference
13.396	45			
13.262	50			
13.139	55			
13.017	60			
		$pK = 4787.3/T + 7.1321 \log T + 0.010365T - 22.801$ (T in ^{O}K) Thermodynamic quantities are calculated from the results		
14.955	0	For $I = 0$; from calorimetric measurements on electrolyte solutions		A9
14.534	10			
14.161	20			
13.999	25			
13.833	30			
13.533	40			
13.262	50			
13.015	60			
12.800	70			
12.598	80			
12.422	90			
12.259	100			
12.126	110			
12.002	120			
11.907	130			
14.926	0	$I = 0.0001$ to 1.5; potentials corrected to $I = 0$ using equations of the form, $E_{corr} = E_{obs} + \alpha I^{1/2} - \beta I$	E3ah	B82
14.222	18			
13.980	25			
13.590	37			
13.05	60	Predicted from thermodynamic data, taking pK_w = 13.997 at 25°		C36
12.21	100			
11.65	150			

11.30	200			
11.18	250			
11.19	300			
11.33	350			
12.32	100	From measurements of the degree of hydrolysis of	Cl	N46
11.65	156	ammonium acetate		
11.34	218			
11.77	306			
13.907	25	I = 0.1, taking pK_w = 13.997 at 1 atmosphere;	Elcg	H16
		250 atmospheres		
13.824		500 atmospheres		
13.747		750 atmospheres		
13.667		1000 atmospheres		
13.585		1250 atmospheres		
13.524		1500 atmospheres		
13.449		1750 atmospheres		
13.394		2000 atmospheres		
		For predictions of pK_w from 350-700° at superheated-steam densities of 0.3 to 0.7 $g.cm^{-3}$., see F40		
		Ref. H56 gives an equation for the variation of pK_w with temperature from 0° to 306°		
		For values of pK_w in H_2O/D_2O mixtures, see S8		
29		pK of OH^-; theoretical prediction		S48
23.2		pK of OH^-, assuming the difference between pK_1 and pK_2 of H_2O is the same as for H_2S		B80
14.747	5	Molal scale	Elch	B54
14.551	10	Extrapolated to I = 0		
14.360	15			
14.182	20			
14.016	25			
13.850	30			

Name, Formula and pK value	T(^{o}C)	Remarks	Methods	Reference
13.700	35			
13.548	40			
13.414	45			
13.280	50			
12.979	65			
12.771	75			
12.666	80			
12.562	85			
13.236	51.0	From specific conductivity of pure water	C1	B69
13.007	60.9			
12.821	70.5			
12.668	80.2			
12.520	89.9			
12.271	99.3			
12.117	108.6			
12.036	118.0			
11.926	127.3			
11.831	136.7			
11.793	146.0			
11.688	155.4			
11.620	164.7			
11.571	174.2			
11.501	183.6			
11.451	193.0			
11.408	202.5			
11.400	212.1			
11.387	221.7			
11.383	231.4			
11.383	241.2			

11.392	251.0			
11.417	261.0			
11.451	271.0			
		$-\log K_w = 4387.93/T - 5.64327 + 0.0165276\,T$		
14.941	0	Self-dissociation of water in 0.02-2.7 m KCl; I = 0.1 to 3.0, extrapolated to I = 0	E2b	S130
13.993	25			
13.272	50			
12.709	75			
12.264	100			
11.914	125			
11.642	150			
11.441	175			
11.302	200			
11.222	225			
11.196	250			
11.224	275			
11.301	300			
13.997	25	molal scale; from measurements in solutions of ammonia and acetic acid	C1	F25
13.262	50			
12.697	75			
12.26	100			
11.91	125			
11.64	150			
11.42	175			
11.26	200			
11.14	225			
11.05	250			
11.01	275			

Name, Formula and pK value	T(^{o}C)	Remarks	Methods	Reference
11.42	300			
12.17	100	$\underline{I}$ = 0.02-0.10, extrapolated to $\underline{I}$ = 0; molal scale	E3ah	D56
11.96	125			
11/72	150			
11.43	175			
11.27	200			
13.997	25	$p\underline{K}_w$; molal scale	C	F24
12.254	100			
11.634	150			
11.254	200			
11.049	250			
11.034	300			
11.422	350			
13.95	25	$\underline{I}$ = 1.5($NaClO_4$)	E,g	B152
14.97	0	$\underline{I}$ = 3($NaClO_4$)		
14.09	25			
13.66	40			
13.10	60			
12.83	25	c = 0.5 M Na_2SO_4		
20.8		$p\underline{K}$ of OH^-, assuming the difference between $p\underline{K}_1$ and $p\underline{K}_2$ of H_2O is the same as for H_2S, using more recent values for H_2S, H_2Se and H_2Te		E3
		For the basic $p\underline{K}$ of H_2O in concentrated H_2SO_4, see G35		
		For $p\underline{K}_w$ from 0-1000° and for 1-10000 bars, see M34		
		For $p\underline{K}_w$ to 800° and 4000 bars, see Q5		
		For $p\underline{K}_w$ to 1000° and 100 kilobars, see H83, H84		
		For $p\underline{K}_w$ to 350°, see F25		
		For $p\underline{K}_w$ in NaCl media (1-3 $\underline{m}$ NaCl), from 25° to 295°,		

		see B157. For 1-6$\underline{m}$ NaCl at 25°, see F24.		
		For $p\underline{K}_w$ in DMSO/water mixtures, see F22a		
		Other measurements: A15, A19, B31, B69, B130, C6, D32, F23, H35, H38, H40, H60, H61, I3, I9, L2, L31, M5, M63, N14, N15, N36, O12, P30, R25, W24, W45		

262. Xenon trioxide, XeO_3

10.5	25	$p\underline{K}$ for XeO_3.aq. $\rightleftharpoons HXeO_4^- + H^+$; $\underline{I}$ = 0.5($NaClO_4$)	E3bg	A43
10.8		$\underline{I}$ = 0.1		
10.1		$p\underline{K}$ of perxenic acid, H_4XeO_6; $\underline{I}$ = 0.1		

263. (Aquo) Ytterbium(III) ion, Yb^{3+}

7.92	25	$p\underline{K}_a$ for hydrolysis of Yb^{3+}; titration of 0.004-0.009 M $Yb(ClO_4)_3$ with 0.02 M $Ba(OH)_2$; $\underline{I}$ = 0.3($NaClO_4$)	E3b	F49
8.03	25	ditto, using 0.02 M NaOH		
8.01	25	$\underline{I}$ = 0	E3bg	U2

264. (Aquo) Yttrium(III) ion, Y^{3+}

9.10	25	$p\underline{K}$ for hydrolysis of Y^{3+}; $\underline{I}$ = 3($LiClO_4$); c = 0.01-1 M $Y(ClO_4)_3$; $Y_2(OH)_2^{4+}$ (log $\underline{K}$ = − 14.28) and $Y_3(OH)_5^{4+}$ (log $\underline{K}$ = − 33.8) are also formed	E3ag,quin	B62
8.34	25	$p\underline{K}_a$ for hydrolysis of Y^{3+}; titration of 0.004-0.009 M $Y(ClO_4)_3$ with 0.02 M $Ba(OH)_2$; $\underline{I}$ = 0.3($NaClO_4$)	E3b	F49
8.04	25	$\underline{I}$ = 0	E3bg	U2
16.8	25	−log $\underline{K}$ for $Y(OH)_2^+$; $\underline{I}$ = 3($LiClO_4$)	E3bg	A26
14.04		−log β_{22} for $Y_2(OH)_2^{4+}$		
17.0	25	−log $\underline{K}$ for $Y^{3+} + 2D_2O \rightleftharpoons Y(OD)_2^+ + 2D^+$; $\underline{I}$ = 3($LiClO_4$) in D_2O	E3bg	A27
14.75		−log β_{22} for $2Y^{3+} + 2D_2O \rightleftharpoons Y_2(OD)_2^{4+} + 2D^+$; $\underline{I}$ =		

Name, Formula and pK value	T(^{o}C)	Remarks	Methods	Reference
		3($LiClO_4$) in D_2O		
265. (Aquo) Zinc ion, Zn^{2+}				
9.30	15	p$\underline{K}$ for hydrolysis of Zn^{2+} to $ZnOH^+$; $\underline{I}$ = 0.0015 to 0.04 (KNO_3); extrapolated to $\underline{I}$ = 0	E3bg	P36
9.15	20			
8.96	25			
8.79	30			
8.62	36			
8.46	42			
8.7	30	p$\underline{K}$ for hydrolysis of Zn^{2+}; $\underline{I}$ = 0.1(KCl)	E3bg	C14
9.01	25	p$\underline{K}$ for hydrolysis of Zn^{2+}; $\underline{I}$ = 2(KCl); c = 0.1 M $ZnCl_2$, also −log $\underline{K}$ = 7.20 for $2Zn^{2+} + H_2O \rightleftharpoons Zn_2OH^{3+} + H^+$	E3ag	S35
9.12	25	ditto; $\underline{I}$ = 2(NaCl); −log $\underline{K}$ = 7.48 for $2Zn^{2+} + H_2O \rightleftharpoons Zn_2OH^{3+} + H^+$		
6.31 4.88 3.12 3.39	25	Stepwise p$\underline{K}_b$ values for Zn^{2+}, $\underline{I}$ = 1($NaClO_4$)	SOLY	G68
9.05	25		E3bg	D72
7.87	100	p$\underline{K}$ for hydrolysis of Zn^{2+}; c = 0.02 M $Zn(NO_3)_2$	KIN	K108
5.7		p$\underline{K}_b$ for $ZnOH^+$; c = 0.1 M $ZnSO_4$	E3bg	S44
8.73		p$\underline{K}$ for $Zn(OH)_2 \rightleftharpoons Zn(OH)_3^- + H^+$; $\underline{I}$ = 3($NaClO_4$)	DISTRIB	S57
9.89		p$\underline{K}$ for $Zn(OH)_3^- \rightleftharpoons Zn(OH)_4^{2-} + H^+$; $\underline{I}$ = 3($NaClO_4$); also log $\underline{K}$ = − 20.10 for $[Zn(OH)_2][H^+]^2/[Zn^{2+}]$		
8.7	25	log $\underline{K}$ for $2Zn^{2+} + OH^- \rightleftharpoons Zn_2OH^3$; $\underline{I}$ = 3($LiClO_4$); c = 0.25 − 1.45M Zn^{2+}		B58
16.84 16.91	0	log $\underline{K}$ for $Zn^{2+} + 3OH^- \rightleftharpoons Zn(OH)_3^-$, and log $\underline{K}$ for $Zn^{2+} + 4OH^- \rightleftharpoons Zn(OH)_4^{2-}$	POLAROG	K75
15.86 15.95	20			
15.45 15.55	30			
15.15	25	log $\underline{K}$ for $Zn^{2+} + 4OH^- \rightleftharpoons Zn(OH)_4^{2-}$; zinc electrode potential measurements		D54

16.9		log $\underline{K}$ for $Zn^{2+} + 4OH^- \rightleftharpoons Zn(OH)_4^{2-}$	POLAROG	S107
15.45	25	log $\underline{K}$ for $Zn^{2+} + 4OH^- \rightleftharpoons Zn(OH)_4^{2-}$; zinc electrode potential measurements		D53
15.3	25	log $\underline{K}$ for $Zn^{2+} + 4OH^- \rightleftharpoons Zn(OH)_4^{2-}$; $\underline{I}$ = 2(KCl)	POLAROG	M48a
14.5	18	log $\underline{K}$ for $Zn^{2+} + 4OH^- \rightleftharpoons Zn(OH)_4^{2-}$; zinc electrode potential measurements		S106
13.35	25	log $\underline{K}$ for $Zn^{2+} + 4OH^- \rightleftharpoons Zn(OH)_4^{2-}$; $\underline{I}$ = 3(NaCl); Zn-Hg electrode		S36
26.77	25	log $\underline{K}$ for $2Zn^{2+} + 6OH^- \rightleftharpoons Zn_2(OH)_6^{2-}$		
15.04	20	log $\underline{K}$ for $Zn^{2+} + 4OH^- \rightleftharpoons Zn(OH)_4^{2-}$; Zn-Hg electrode		B48
16.08	20	log $\underline{K}$ for $Zn^{2+} + 3OH^- \rightleftharpoons Zn(OH)_3^-$		
13.58	20	log $\underline{K}$ for $Zn^{2+} + 3OH^- \rightleftharpoons Zn(OH)_3^-$; $\underline{I}$ = 0.1	DISTRIB	B86
9.01	25	p$\underline{K}$ for hydrolysis of Zn^{2+}; $\underline{I}$ = 2(KCl)	E3bg	S38,S104
7.20		–log $\underline{K}$ for $2Zn^{2+} + H_2O \rightleftharpoons Zn_2OH^{3+} + H^+$; $\underline{I}$ = 2(KCl)		
9.11		p$\underline{K}$ for hydrolysis of Zn^{2+}; $\underline{I}$ = 2(NaCl)		
7.49		–log $\underline{K}$ for $2Zn^{2+} + H_2O \rightleftharpoons Zn_2OH^{3+} + H^+$; $\underline{I}$ = 2(NaCl)		
9.25	25	p$\underline{K}$ for hydrolysis of Zn^{2+}; $\underline{I}$ = 3(NaCl)		S37
9.26		$\underline{I}$ = 3(KCl)		
7.50	25	–log $\underline{K}$ for $2Zn^{2+} + H_2O \rightleftharpoons Zn_2OH^+ + H^+$; $\underline{I}$ = 3(NaCl)		
7.47		$\underline{I}$ = 3(KCl)		
10.4	25	p$\underline{K}$ for Zn^{2+}; $\underline{I}$ = 3($NaClO_4$); c = 0.6 to 1.25 M Zn^{2+}	E3bg	B146,B148
9.5	60			
8.7	100			
8.72	25	–log $\underline{K}$ for $2Zn^{2+} + H_2O \rightleftharpoons Zn_2OH^+ + H^+$; $\underline{I}$ = 3($NaClO_4$)		
7.62	60			
6.50	100			
8.6	25	–log $\underline{K}$ for $2Zn^{2+} + H_2O \rightleftharpoons Zn_2OH^+ + H^+$; $\underline{I}$ = 3($NaClO_4$); c = 0.2-1.5 M in Zn^{2+}	E3bg	Z4
26.00	25	–log $\underline{K}$ for $4Zn^{2+} + 4H_2O \rightleftharpoons Zn_4(OH)_4^{4+} + 4H^+$ Polynuclear hydrolysed species in zinc solutions at		

Name, Formula and pK value	T(°C)	Remarks	Methods	Reference
		pH above 8 are postulated from coagulation studies		M48
		Other measurements: A5, B134, D18, D37, F52, H5, K12, K62, K109, N21, P65, Q2, S34, S90, W34		
266. (Aquo) Zirconium(IV) ion, Zr^{4+}				
−0.32 0.06 0.35 0.64	25	Successive p$\underline{K}$ values for the hydrolysis of Zr^{4+}; $\underline{I}$ = 1($HClO_4$); low Zr^{4+} concentrations; at higher concentrations polymers are also formed	DISTRIB	P40
0.22 0.62 1.05 1.17	25	Successive p$\underline{K}$ values for the hydrolysis of Zr^{4+}; $\underline{I}$ = 2(HCl,HNO_3)	DISTRIB	S92
14.58 14.80 14.34 14.13	25	Successive p$\underline{K}_b$ values for hydrolysis of Zr^{4+}; $\underline{I}$ = 1($LiClO_4$,$NaClO_4$); tracer concentrations of Zr(IV)	DISTRIB	S93
−0.30	25	p$\underline{K}$ for $Zr^{4+} + H_2O \rightleftharpoons ZrOH^{3+} + H^+$	DISTRIB	N43
+0.65	25	p$\underline{K}$ for $Zr^{4+} + H_2O \rightleftharpoons ZrOH^{3+} + H^+$; based on effect of pH on complex formation by fluoride ion, using a fluoride membrane electrode; $\underline{I}$ = 4($Na^+ + H^+$)ClO_4	ION	
5.60		−log $\underline{K}$ for $3ZrOH^{3+} + (4-n)H_2O \rightleftharpoons Zr_3O_n(OH)^{5+}_{(7-n)} + 4H^+$; $\underline{I}$ = 3.5; c = 0.02 M	O8	T26

REFERENCES

A

A1 G. A. Abbott and W.C. Bray J. Am. Chem. Soc. 31 729 (1909)

A2 E. Abel, E. Bratu and O. Redlich Z. Physik. Chem. A173 353 (1935)

A3 E. Abel and J. Proisl Monatsh. 72 1 (1938)

A4 E. Abel, O. Redlich and P. Hersch Z. Physik. Chem. A170 112 (1934)

A5 F. Achenza Ann. Chim. (Rome) 48 565 (1958)

A6 F. Achenza Ann. Chim. (Rome) 49 624 (1959)

A7 F. Achenza Red. Seminar. Fac. Sci. Univ. Cagliari 29 52 (1959); CA 54 21954 (1960)

A8 F. Achenza Ann. Chim. (Rome) 54 240 (1964)

A9 T. Ackermann Z. Elektrochem. 62 411 (1958)

A10 M. N. Ackermann and R. E. Powell Inorg. Chem. 5 1334 (1966)

A11 L. P. Adamovich and G. S. Shupenko Ukrain. Khim. Zhur. 25 155 (1959); CA 53 18601 (1959)

A12 M. Adhikari, D. Ganguli, G. Biswas and D. Ghosh J. Indian Chem. Soc. 46 1131 (1969)

A13 V. N. Afanasev, V. A. Shormanov and G. A. Gestov Tr. Ivanov. Khim-Tekhnol. Inst. 13 36 (1972)

A14 A. L. Agalonova and I. L. Agafonov Zhur. Fiz. Khim. 27 1137 (1953)

A15 A. Agren Acta Chem. Scand. 9 49 (1955)

A16 I. Ahlberg Acta Chem. Scand. 16 887 (1962)

A17 S. Ahrland Acta Chem. Scand. 3 374 (1949)

A18 S. Ahrland and L. Brandt Acta Chem. Scand. 22 1579 (1968)

A19 S. Ahrland and I. Grenthe Acta Chem. Scand. 11 111 (1957)

A20 E. Ahrland, R. Larsson and K. Rosengren Acta Chem. Scand. 10 705 (1956)

A21 S. Akelin and U. Y. Ozer, J. Inorg. Nucl. Chem. 33 4171 (1971)

A22 J. W. Akitt, A. K. Covington, J. G. Freeman and T. H. Lilley Chem. Comm. 1965 349

A23 Yu. V. Alekhin, A. V. Zotov and N. N. Kolpakova, Ocheski Fiz. Khim. Petrol. 6 5 (1977)

A24 I. P. Alimarin, S. A. Kamid and I. V. Puzdrenkova Russ. J. Inorg. Chem. (Engl. trans.) 10 209 (1965)

A25 T. L. Allen and R. M. Keefer J. Am. Chem. Soc. 77 2957 (1955)

A26 T. Amaya, H. Kakihana and M. Maeda Bull. Chem. Soc. Japan

A27 T. Amaya, H. Kakihana and M. Maeda Bull. Chem. Soc. Japan 46 2889 (1973)

A28 A. R. Amell J. Am. Chem. Soc. 78 6234 (1956)

A29 N. I. Ampelogova, Radiokhimiya 17 68 (1975)

A30 G. Anderegg, Helv. Chim. Acta 40 1022 (1957)

A31 G. Anderegg, Helv. Chim. Acta 48 1712 (1965)

A32 G. Anderegg, G. Schwarzenbach, M. Padmoyo and O. F. Borg Helv. Chim. Acta 41 988 (1958)

A33 K. P. Ang J. Chem. Soc. 1959 3822

A34 P. J. Antikainen Suomen Kem. 28B 135 (1955)

A35 P. J. Antikainen Acta Chem. Scand. 10 756 (1956)

A36 P. J. Antikainen Suomen Kem. 30B 123 (1957)

A37 P. J. Antikainen Suomen Kem. 30B 201 (1957)

A38 P. J. Antikainen and D. Dyrssen Acta Chem. Scand. 14 86 (1960)

A39 P. J. Antikainen and V.M.K. Rossi Suomen Kem. 32B 185 (1959)

A40 P. J. Antikainen and K. Tevanen Suomen Kem. 34B 3 (1961)

A41 V. P. Antonovich and V. A. Nazarenko Zhur. Neorg. Khim 13 805 (1968)

A42 V. P. Antonovich, E. M. Nevskaya and E. N. Suvorova Zhur. Neorg. Khim. 22 1278 (1977)

A43 E. H. Appelman and J. G. Malm J. Am. Chem. Soc. 86 2141 (1964)

A44 T. V. Arden J. Chem. Soc. 1951 350

A45 G. P. Arkhipova, I. E. Flis and K. P. Mishchenko Russ. J. Appl. Chem. (Engl. trans.) 37 2275 (1964)

A46 G. P. Arkhipova, K. P. Mishchenko and I. E. Flis Zhur. Priklad. Khim. (Leningrad) 41 1131 (1968)

A47 K. Arndt, Z. Physik. Chem. 45 571 (1903)

A48 R. Arnek, Acta Chem. Scand. 23 1986 (1969)

A49 E. M. Arnett and R. D. Bushick J. Am. Chem. Soc. 86 1564 (1964)

A50 E. M. Arnett and G. W. Mach J. Am. Chem. Soc. 86 2671 (1964)

A51 E. M. Arnett and G. W. Mach J. Am. Chem. Soc. 88 1177 (1966)

A52 K. G. Ashurst and W. C. Higginson J. Chem. Soc. 1956 343

A53 R. W. Asmusson and L. T. Muus Trans. Danish Acad. Tech. Sci. 1946 No. 1, 3

A54 L. F. Audrieth and S. F. West J. Am. Chem. Soc. 77 5000 (1955)

A55 F. Z. Auerbach Z. Physik. Chem. 49 217 (1904)

A56 M. Auméras J. Chim. Phys. 25 300 (1928)

A57 J. Aveston Inorg. Chem. 3 981 (1964)

A58 J. Aveston J. Chem. Soc. 1965 4438

A59 J. Aveston J. Chem. Soc., A 1966 1599

A60 J. Aveston, E. W. Anacker and J. S. Johnson Inorg. Chem. 3 735 (1964)

A61 S. Aybar Comm. Fac. Sci. Univ. Ankara 5B 22 (1954); CA 49 9362 (1955)

A62 S. Aybar Comm. Fac. Sci. Univ. Ankara 10B 44 (1962); CA 59 13397 (1963)

B

B1 A. K. Babko, V. V. Lukachina and B. I. Nabivanets Russ. J. Inorg. Chem. (Engl. trans.) 8 957 (1963)

B2 C. F. Baes J. Am. Chem. Soc. 79 5611 (1957)

B3 C. F. Baes and N.J. Meyer Inorg. Chem. 1 780 (1962)

B4 C. F. Baes, N. J. Meyer and C. E. Roberts Inorg. Chem. 4 518 (1965)

B5 N. Bailey, A. Carrington, K. A. K. Lott and M. C. R. Symons J. Chem. Soc. 1960 290

B6 T. A. Bak and E. L. Prestgaard Acta Chem. Scand. 11 901 (1957)

B7 F. B. Baker and T. W. Newton J. Phys. Chem. 61 381 (1957)

B8 F. B. Baker, T. W. Newton and M. Kahn J. Phys. Chem. 64 109 (1960)

B9 W. G. Baldwin and L. G. Sillen, Arkiv. Kemi 31 391 (1969)

B10 W. G. Baldwin and G. Wiese, Arkiv. Kemi 31 419 (1969)

B11 D. L. Ball and J. O. Edwards J. Am. Chem. Soc. 78 1125 (1956)

B12 W. G. Barb, J. H. Baxendale, P. George and K. R. Hargrave Trans. Faraday Soc. 47 591 (1951)

B13 L. Barcza and L. G. Sillén Acta Chem. Scand. 25 1250 (1971)

B14 J. Barr, R. J. Gillespie and E. A. Robinson, Canad. J. Chem. 39 1266 (1961)

B15 M. Bartusck and L. Sommer, Z. Phys. Chem. (Leipzig) 226 309 (1964)

B16 K. N. Bascombe and R. P. Bell J. Chem. Soc. 1959 1096

B17 S. J. Bass, R. J. Gillespie and E. A. Robinson J. Chem. Soc. 1960 821

B18 R. G. Bates J. Res. Natl. Bur. Std. 47 127 (1951)

B19 R. G. Bates and S. F. Acree J. Res. Natl. Bur. Std. 30 129 (1943)

B20 R. G. Bates and S. F. Acree J. Res. Natl. Bur. Std. 34 373 (1945)

B21 R. G. Bates, V. E. Bower, R. G. Canham and J. E. Prue Trans. Faraday Soc. 55 2062 (1959)

B22 R. G. Bates and R. Gary J. Res. Natl. Bur. Std. 65A 495 (1961)

B23 R. G. Bates and G. D. Pinching J. Res. Natl. Bur. Std. 42 419 (1949)

B24 R. G. Bates and G. D. Pinching J. Am. Chem. Soc. 72 1393 (1950)

B25 C.J. Battaglia and J. O. Edwards Inorg. Chem. 4 552 (1965)

B26 E. Bauer Z. Physik. Chem. 56 215 (1906)

B27 D. Bauer and M. Bouchet C.R. Acad. Sci. (Paris) 275 21 (1972)

B28 E. W. Bauman J. Inorg. Nucl. Chem. 31 3155 (1969)

B29 N. V. Bausova and L. L. Manakova, Zhur. Neorg. Khim. 19 1213 (1974)

B30 J. C. Bavay, G. Nowogrocki and G. Tridot Bull. Soc. Chim. France 1967 2026

B31 H. T. Beans and E. T. Oakes J. Am. Chem. Soc. 42 2116 (1920)

B32 K. Beck and P. Stegmüller Arb. Kaiser Gesundh. 34 446 (1910)

B33 M. A. Beg, Kabir-ud-Din and R. A. Khan, Austral. J. Chem. 26 671 (1973)

B34 R. P. Bell The Proton in Chemistry, Chap. 7 Cornell Univ. Press, New York (1959)

B35 R. P. Bell, K. N. Bascombe and J. C. McCoubrey J. Chem. Soc. 1956 1286

B36 R. P. Bell, A. L. Dowding and J.A. Noble J. Chem. Soc. 1955 3106

B37 R. P. Bell and E. Gelles J. Chem. Soc. 1951 2734

B38 R. P. Bell and J. H. B. George Trans. Faraday Soc. 49 619 (1953)

B39 R. P. Bell and M. H. Panckhurst J. Chem. Soc. 1956 2836

B40 R. P. Bell and J. E. Prue J. Chem. Soc. 1949 362

B41 R. P. Bell and G. M. Waind J. Chem. Soc. 1950 1979

B42 E. A. Belousov and V.M. Ivanov, Zhur. Fiz. Khim. 51 529 (1977)

B43 T. Beltran and C. Mateo Anales Real. Soc. Espan. Fiz. Quim. 61B, 12 1219 (1965)

B44 M. Beran Coll. Czech. Chem. Comm. 32 1368 (1967)

B45 C. Berecki-Biedermann Arkiv. Kemi. 9 175 (1956)

B46 D. Berg and A. Patterson J. Am. Chem. Soc. 75 5197 (1953)

B47 D. Berg and A. Patterson J. Am. Chem. Soc. 75 5731 (1953)

B48 P. Bernheim and M. Quentin Compt. Rend. 230 388 (1950)

B49 F. Bertin, G. Thomas and J. C. Merlin Compt. Rend. 260 1670 (1965)

B50 J. Bessièrre Anal. Chim. Acta 52 55 (1970)

B51 R. H. Betts Canad. J. Chem. **33** 1775 (1955)

B52 J. Beukenkamp and K. D. Herrington J. Am. Chem. Soc. **82** 3025 (1960)

B53 J. Beukenkamp, W. Rieman and S. Lindenbaum Anal. Chem. **26** 505 (1954)

B54 C. P. Bezboruah, M. Camoes, A. Covington and J. Dobson J. Chem. Soc., Faraday Trans. I **69** 949 (1973)

B55 G. Bhat and R. S. Subrahamanya, Proc. Indian Acad. Sci., Sec. A, **73** 157 (1971)

B56 T. F. Bidleman, Anal. Chim. Acta **56** 221 (1971)

B57 G. Biedermann Rec. Trav. Chim. **73** 716 (1956); Arkiv Kemi **9** 277 (1956)

B58 G. Biedermann Proc. 7th Int. Conf. Coordin. Chem. Stockholm **1962** 159

B59 G. Biedermann Svensk. Kem. Tidskr. **76** 362 (1964)

B60 G. Biedermann and L. Ciavatta Acta Chem. Scand. **15** 1347 (1961)

B61 G. Biedermann and L. Ciavatta Acta Chem. Scand. **16** 2221 (1962)

B62 G. Biedermann and L. Ciavatta Arkiv Kemi **22** 253 (1964)

B63 G. Biedermann and S. Hietanen Acta Chem. Scand. **14** 711 (1960)

B64 G. Biedermann, M. Kilpatrick, I. Pokras and L. G. Sillén Acta Chem. Scand. **10** 1327 (1956)

B65 G. Biedermann, N. C. Li and J. Yu Acta Chem. Scand. **15** 555 (1961)

B66 G. Biedermann and L. Newman Arkiv Kemi **22** 303 (1964)

B67 G. Biedermann and R. Palombari, Acta Chem. Scand., A, **32** 381 (1978)

B68 G. Biedermann and L. G. Sillén Acta Chem. Scand. **14** 717 (1960)

B69 G. J. Bignold, A. D. Brewer and B. Hearn Trans. Faraday Soc. **67** 2419 (1971)

B70 H. Bilinski and N. Ingri Acta Chem. Scand. **21** 2503 (1967)

B71 C. Birraux, J. C. Landry and W. Haerdi Anal. Chim. Acta **93** 281 (1977)

B72 E. A. Biryuk and V. A. Nazarenko Zhur. Neorg. Khim. **18** 2964 (1973)

B73 E. A. Biryuk, V. A. Nazarenko and R. V. Ravitskaya Zhur. Neorg. Khim. **14** 565 (1968)

B74 V. P. Biryukov and E. Sh. Ganelina, Zhur. Neorg. Khim. **16** 600 (1971)

B75 H. Bilinski and N. Ingri Acta Chem. Scand. **21** 2503 (1967)

B76 T. C. Bissot, R. W. Parry and D. H. Campbell J. Am. Chem. Soc. **79** 796 (1957)

B77 N. Bjerrum Kgl. Danske Videnskab Selskab Skrifter, Nat.-mat. Afd. **4** 1 (1906)

B78 N. Bjerrum Z. Physik. Chem. **59** 336 (1907)

B79 N. Bjerrum Z. Physik. Chem. **73** 724 (1910)

B80 N. Bjerrum Z. Physik. Chem. **106** 219 (1923)

B81 J. Bjerrum Diss., Copenhagen, **1941**

B82 N. Bjerrum and A. Unmack Kgl. Danske Videnskab Selskab, Mat-fys Medd. **9** No. 1 (1929)

B83 E. Blanc J. Chim. Phys. **18** 28 (1920)

B84 A. A. Blanchard Z. Physik. Chem. **41** 681 (1902)

B85 B. Blaser and K. H. Worms Z. Anorg. Allgem. Chem. **301** 18 (1959)

B86 H. Bode Z. Anorg. Allgem. Chem. **317** 3 (1962)

B87 R. H. Bogue J. Am. Chem. Soc. **42** 2575 (1920)

B88 J. A. Bolzan Rev. Fac. Cienc. Quim., Univ. Nacl. La Plata **33** 67 (1960); CA **58** 3091 (1963)

B89 J. A. Bolzan and A. J. Arvia Electrochim. Acta 7 589 (1962)

B90 J. A. Bolzan and A. J. Arvia Electrochim. Acta 8 375 (1963)

B91 J. A. Bolzan, E. A. Jauregui and A. J. Arvia Electrochim. Acta 8 841 (1963)

B92 J. A. Bolzan, J. J. Podesta and A. J. Arvia Anales Asoc. Quim. Argentine 51 43 (1963)

B93 T. G. Bonner and J. C. Lockhart J. Chem. Soc., 1957 2840

B94 O. Borgen, M. R. Mahmoud and I. Skauvik Acta Chem. Scand. A31 329 (1977)

B95 V. A. Borgoyakov and E. Sh. Ganelina Izv. Vysh. Ucheb. Zaved. Khim., Khim Tekhnol. 16 705 (1973)

B96 E. Bottari and L. Ciavatta J. Inorg. Nucl. Chem. 27 133 (1965)

B97 J. H. Boughton Diss. Abstr. 26 5710 (1966)

B98 J. H. Boughton and R. N. Keller J. Inorg. Nucl. Chem. 28 2851 (1966)

B99 K. Bowden Canad. J. Chem. 43 2624 (1965)

B100 R. H. Boyd J. Am. Chem. Soc. 83 4288 (1961)

B101 R. H. Boyd J. Am. Chem. Soc. 85 1555 (1963)

B102 G. E. K. Branch, D. L. Yabroff and B. Bettman J. Am. Chem. Soc. 56 937 (1934)

B103 J. C. D. Brand J. Chem. Soc. 1950 997

B104 J. C. D. Brand, W. C. Horning and M. B. Thornley J. Chem. Soc. 1952 1374

B105 J. C. D. Brand, A. W. P. Jarvie and W. C. Horning J. Chem. Soc. 1959 3844

B106 E. A. Braude J. Chem. Soc. 1948 1971

B107 W. C. Bray and A. V. Hershey J. Am. Chem. Soc.56 1889 (1934)

B108 W. C. Bray and H. A. Liebhafsky J. Am. Chem. Soc. 57 51 (1935)

B109 G. Bredig Z. Physik Chem. 13 289 (1894)

B110 R. Brinkman, R. Margaria and F. J. W. Roughton Proc. Roy. Soc. (London) 232A 65 (1933)

B111 F. Brito Anales Real Soc. Espan. Fis. Quim. (Madrid), Ser. B, 62 128 (1966); CA 65 3075 (1966)

B112 F. Brito Anales Real Soc. Espan. Fis. Quim. (Madrid), Ser. B, 62 193 (1966); CA 65 3082 (1966)

B113 F. Brito Anales Real Soc. Espan. Fis. Quim. (Madrid), Ser. B, 62 197 (1966); CA 65 3075 (1966)

B114 F. Brito, N. Ingri and L. G. Sillén Acta Chem. Scand. 18 1557 (1964)

B115 H. T. S. Britton J. Chem. Soc. 125 1572 (1924)

B116 H. T. S. Britton J. Chem. Soc. 129 614 (1927)

B117 H. T. S. Britton and H. G. Britton J. Chem. Soc. 1952 3892

B118 H. T. S. Britton and E. N. Dodd J. Chem. Soc. 1931 2332

B119 H. T. S. Britton and E. N. Dodd Trans. Faraday Soc. 29 537 (1933)

B120 H. T. S. Britton and P. Jackson J. Chem. Soc. 1934 1048

B121 H. T. S. Britton and R. A. Robinson J. Chem. Soc. 1931 458

B122 H. T. S. Britton and R. A. Robinson Trans. Faraday Soc. 28 531 (1932)

B123 A. Brodskii and L. V. Sulima Doklady Akad. Nauk. S.S.S.R. 85 1277 (1952)
B124 H. H. Broene and T. De Vries J. Am. Chem. Soc. 69 1644 (1947)
B125 S. Broersma J. Chem. Phys. 26 1405 (1957)
B126 J. N. Brönsted and C. V. King J. Am. Chem. Soc. 49 193 (1927)
B127 J. N. Brönsted and K. Pedersen Z. Physik. Chem. 108 185 (1924)
B128 J. N. Brönsted and K. Volqvartz Z. Physik. Chem. 134 97 (1928)
B129 C. Brosset Naturwiss. 29 455 (1941)
B130 C. Brosset Acta Chem. Scand. 6 910 (1952)
B131 C. Brosset, G. Biedermann and L. G. Sillén Acta Chem. Scand. 8 1917 (1954)
B132 C. Brosset and B. Gustaver Svensk. Kem. Tidskr. 54 155 (1942)
B133 C. Brosset and U. Wahlberg Svensk. Kem. Tidskr. 55 335 (1943)
B134 H. F. Brown and J. A. Cranston J. Chem. Soc. 1940 578
B135 S. Bruckenstein and D. C. Nelson J. Chem. Eng. Data 6 605 (1961)
B136 T. C. Bruice and S. J. Benkovic J. Am. Chem. Soc. 86 418 (1964)
B137 T. C. Bruice and J. J. Bruno J. Am. Chem. Soc. 83 3494 (1961)
B138 S. Bruner Z. Electrochem. 19 861 (1913)
B139 J. Buchanan and S. D. Hamann Trans. Faraday Soc. 49 1425 (1953)
B140 J. R. Buchholz and R. E. Powell J. Am. Chem. Soc. 85 509 (1963)
B141 G. T. Buist, W. C. P. Hipperson and J. D. Lewis J. Chem. Soc. A 1969 307
B142 G. T. Buist and J. D. Lewis Chem. Comm. 1965 66
B143 D. Bunn, F.S. Dainton and S. Duckworth Trans. Faraday Soc. 57 1131 (1961)
B144 K. A. Burkov, E. A. Bus'ko, L. A. Garmash and G.V. Khonin, Zhur. Neorg. Khim 23 1767 (1978)
B145 K. A. Burkov, E. A. Bus'ko and N. I. Zinevich Vestn. Leningrad. Univ. 19, Khim. 1978 144
B146 K. A. Burkov and L. A. Garmash Zhur. Neorg. Khim. 22 536 (1977)
B147 K. A. Burkov, L. A. Garmash and L. S. Lilich Vestn. Leningrad. Univ., Fiz. Khim. 1977 83
B148 K. A. Burkov, L. A. Garmash and L. S. Lilich Zhur. Neorg. Khim. 23 3193 (1978)
B149 K. A. Burkov and L. V. Ivanova, Vestn. Leningrad. Univ., Fiz. Khim 1966 120
B150 K. A. Burkov and N. V. Kemenetskaya Vestn. Leningrad. Univ., Fiz. Khim. 1978 133
B151 K. A. Burkov and L. S. Lilich Vestn. Leningrad. Univ. Fiz. Khim 1965 103
B152 K. A. Burkov, L. S. Lilich and Nguyen Dinh Ngo, Izv. Vyssh. Uchebn. Zaved. Khim., Khim. Tekhnol. 18 181 (1975)
B153 K. A. Burkov, L. S. Lilich, Nguyen Dinh Ngo and A. Yu. Smirnov Zhur. Neorg. Khim. 18 1513 (1973)
B154 K. A. Burkov, L. S. Lilich and L. G. Sillén Acta Chem. Scand. 19 14 (1965)
B155 K. A. Burkov, N. L. Zinovich and L. S. Lilich Izv. Vyssh. Uchebn. Zaved. Khim., Khim. Tekhnol. 13 1250 (1974)

B156 E. A. Burns and F. D. Chang J. Phys. Chem. 63 1314 (1959)
B157 R. H. Busey and R. E. Mesmer J. Soln. Chem. 5 147 (1976)
B158 R. H. Busey and R. E. Mesmer Inorg. Chem. 16 2444 (1977)
B159 F. J. J. Buytendyk, R. Brinkman and H. W. Mook Biochem. J. 21 576 (1927)
B160 V. Ya. Bytenskii and E. S. Sorohin Zhur. Prikl. Khim. (Leningrad) 45 2130 (1972)

C

C1 W. B. Campbell and O. Maas Canad. J. Res. 2 42 (1930)
C2 J. P. Candlin and R. G. Wilkins J. Chem. Soc. 1960 4236
C3 J. P. Candlin and R. G. Wilkins J. Am. Chem. Soc. 87 1490 (1965)
C4 R. Caramazza Gazz. Chim. Ital. 87 1507 (1957)
C5 R. Caramazza Gazz. Chim. Ital. 88 308 (1958)
C6 B. Carell and A. Olin Acta Chem. Scand. 14 1999 (1960)
C7 B. Carell and A. Olin Acta Chem. Scand. 15 727 (1961)
C8 B. Carell and A. Olin Acta Chem. Scand. 15 1875 (1961)
C9 G. Carpéni Bull. Soc. Chim. France 1943 629
C10 E. Carrière and H. Guiter Bull. Soc. Chim. France 1947 267
C11 R. L. Carroll and R. E. Mesmer Inorg. Chem. 6 1137 (1967)
C12 A. Cassel, L. Magon, P. Portanova and F. Tondello Radiochim. Acta 17 28 (1972)
C13 R. Cernatescu and A. Mayer Z. Physik. Chem. A160 305 (1932)
C14 S. Chaberek, R. C. Courtney and A.E. Martell J. Am. Chem. Soc. 74 5057 (1952)
C15 S. Chakrabarti J. Indian Chem. Soc. 44 554 (1967)
C16 S. Chakrabarti and S. Aditya J. Indian Chem. Soc. 48 493 (1971)
C17 L. Chambers J. Cell. Physiol. 68 306 (1966)
C17a J. D. Chanley and E. Feageson J. Am. Chem. Soc. 85 1181 (1963)
C18 R. M. Chapin J. Am. Chem. Soc. 56 2211 (1934)
C19 B. Charreton Comp. Rend. 244 1208 (1957)
C20 F. Chauveau Compt. Rend. 247 1120 (1958)
C21 F. Chauveau Bull. Soc. Chim. France 1960 810
C22 F. Chauveau, P. Souchay and R. Schael Bull. Soc. Chim. France 1959 1190
C23 V. Chavane Ann. Chim. (Paris) [12] 4 383 (1949)
C24 H. Chen and D. E. Irish J. Phys. Chem. 75 2673 2681 (1971)
C25 M. Cher and N. Davidson J. Am. Chem. Soc. 77 793 (1955)
C26 Y. T. Chia U.S. Atomic Energy Comm. UCRL-8311 (1958); CA 53 2914 (1959)
C27 C. W. Childs J. Phys. Chem. 73 2956 (1969)
C28 C. W. Childs Inorg. Chem. 9 2465 (1970)
C29 J. Chojnacka Roczniki Chem. 39 161 (1965)
C30 J. Chojnacka Nucleoniki 12 729 (1967)
C31 J. Chojnacka Roczniki Chem. 47 1359 (1973)

C32 V. G. Chukhlantsev Zhur. Fiz. Khim. 33 3 (1959)
C33 L. Ciavatta Arkiv. Kemi 21 129 (1963)
C34 L. Ciavatta and M. Grimaldi Gazz. Chim. Ital. 103 731 (1973)
C35 L. Ciavatta and M. Grimaldi J. Inorg. Nucl. Chem. 37 163 (1975)
C36 J. W. Cobble J. Am. Chem. Soc. 86 5394 (1964)
C37 J. W. Cobble, in Treatise on Analytical Chemistry, editors I. M. Kolthoff and P. J. Elving, Interscience Publ., New York. Part II, Vol. 6, p. 414 (1964)
C38 J. F. Coetzee and G. R. Padmanathan J. Am. Chem. Soc. 87 5005 (1965)
C39 E. J. Cohn J. Am. Chem. Soc. 49 173 (1927)
C40 M. Cola Gazz. Chim. Ital. 90 1037 (1960)
C41 D. L. Cole, E. M. Eyring, D. T. Rampton, A. Silzars and R. P. Jensen J. Phys. Chem. 71 2771 (1967)
C42 M. P. Collados, P. Brito and R. Diaz Cadaviaco Anales Fiz. Quim. B 1967 843
C43 C. A. Colman-Porter and C. B. Monk J. Chem. Soc. 1952 1312
C44 R. E. Connick, L. G. Hepler, Z.Z. Hugus, J. W. Kury, W. M. Latimer and M. S. Tsao J. Am. Chem. Soc. 78 1827 (1956)
C45 R. E. Connick and M. S. Tsao J. Am. Chem. Soc. 76 5311 (1954)
C46 T. J. Conocchioti, G. H. Nancollas and N. Sutin Inorg. Chem. 5 1 (1966)
C47 B. D. Costley and J. P. G. Farr Chem. and Ind. 1968 1435
C48 L. V. Coulter, J. R. Sinclair, A. G. Cole and G. C. Roper J. Am. Chem. Soc. 81 2986 (1959)
C49 R. P. Courgnaud and B. Trémillon Bull. Soc. Chim. France 1965 752
C50 A. K. Covington, N.B.S. Technical Note 400, ed. R. G. Bates, U.S. Nat. Bur. Stands., 1966, p. 51.
C51 A. K. Covington and J. V. Dobson J. Inorg. Nucl. Chem. 27 1435 (1965)
C52 A. K. Covington, J. V. Dobson and K. V. Srinivasan J. Chem. Soc. Faraday Trans. I 69 94 (1973)
C53 A. K. Covington, J. V. Dobson and W. F. K. Wynne-Jones Trans. Faraday Soc. 61 2057 (1965)
C54 A. K. Covington and R. A. Matheson J. Solution Chem. 5 781 (1976)
C55 A. K. Covington and K. E. Newman J. Inorg. Nucl. Chem. 35 3257 (1973)
C56 A. K. Covington, R. A. Robinson and R. G. Bates J. Phys. Chem. 70 3820 (1966)
C57 A. K. Covington, M. J. Tait and W. F. K. Wynne-Jones Proc. Roy. Soc. A286 235 (1965)
C58 J. A. Cranston and H. F. Brown J. Roy. Tech. Coll. (Glasgow) 4 54 (1937)
C59 C. E. Crouthamel, A. M. Hayes and D. S. Martin J. Am. Chem. Soc. 73 82 (1951)
C60 No entry
C61 C. E. Crouthamel, H. V. Meek, D. S. Martin and C. V. Banks J. Am. Chem. Soc. 71 3031 (1949)
C62 M. M. Crutchfield and J. O. Edwards J. Am. Chem. Soc. 82 3533 (1960)
C63 J. J. Cruywagen Inorg. Chem. 19 552 (1980)

C64 J. Curry and C. L. Hazelton J. Am. Chem. Soc. 60 2773 (1938)
C65 J. Curry and Z. Z. Hugus J. Am. Chem. Soc. 66 653 (1944)
C66 F. Cüta, E. Beránek and J. Pisecký Collection Czech. Chem. Commun. 23 1496 (1958)
C67 F. Cüta, Z. Ksandr and M. Hejunánek Collection Czech. Chem. Commun. 21 1388 (1956)
C68 F. Cüta and B. Polej Chem. Listy 49 473 (1955)
C69 F. Cüta and F. Stráfelda Collection Czech. Chem. Commun. 20 9 (1955)
C70 G. Czapski and B. H. J. Bielski J. Phys. Chem. 67 2180 (1963)
C71 G. Czapski and L. M. Dorfman J. Phys. Chem. 68 1169 (1964)

D

D1 G. Dahlgren and F. A. Long J. Am. Chem. Soc. 82 1303 (1960)
D2 K. Damm and A. Weiss Z. Naturforsch. 10b 534 (1955)
D3 P. R. Danesi Acta Chem. Scand. 21 143 (1967)
D4 P. R. Daris, M. Magini, S. Margherita and G. D. Alessandro Energia Nucleare 15 335 (1968)
D5 L. S. Darken and H. F. Meier, J. Am. Chem. Soc. 64 621 (1942)
D6 U. N. Dash Thermochim. Acta 11 25 (1975)
D7 U. N. Dash Astral. J. Chem. 30 2621 (1977)
D8 H. Dautet and R. Guillaumont Radiochem. Radio and Lett. 8 183 (1971)
D9 Yu. P. Davidov Dokl. Akad. Nauk. Beloruss. S.S.R. 16 524 (1972)
D10 Yu. P. Davidov and V. M. Efremenkov Vestn. Akad. Nauk. Belorus. SSR. Ser. Fiz.-Energ. Nauk. 4 21 (1973)
D11 Yu. P. Davidov and V. M. Efremenkov Radiokhimiya 17 160 (1975)
D12 Yu. P. Davidov and M. A. Grachak Vestn. Akad. Nauk. Belorus. SSR, Ser. Fiz.-Energ. Nauk. 4 31 (1973)
D13 Yu. P. Davidov and G. I. Glazacheva Zhur. Neorg. Khim. 25 1462 (1980)
D14 G. Davidson J. Text. Inst. 24 T185 (1933)
D15 G. F. Davidson J. Chem. Soc. 1954 1649
D16 C. W. Davies J. Chem. Soc. 1938 277
D17 C. W. Davies J. Chem. Soc. 1939 349
D18 C. W. Davies J. Chem. Soc. 1951 1256
D19 G. Davies and A. R. Garafalo Inorg. Chem. 19 3543 (1980)
D20 C. W. Davies and B. E. Hoyle J. Chem. Soc. 1951 233
D21 C. W. Davies and L. J. Hudleston J. Chem. Soc. 125 260 (1924)
D22 C. W. Davies, H. W. Jones and C. B. Monk Trans. Faraday Soc. 48 921 (1952)
D23 C. W. Davies and C. B. Monk J. Chem. Soc. 1949 413
D24 W. G. Davies and J. E. Prue Trans. Faraday Soc. 51 1045 (1955)
D25 D. S. Davis Chem. Met. Eng. 39 615 (1932)
D26 J. A. Davis Thesis, Indiana Univ., 1955
D27 W. Davis and H. J. de Bruin J. Inorg. Nucl. Chem. 26 1069 (1964)
D28 G. G. Davis and W. M. Smith Canad. J. Chem. 40 1836 (1962)
D29 J. G. Dawber J. Chem. Soc. 1965 4111

D30 J. G. Dawber J. Chem. Soc. A **1968** 1532

D31 J. G. Dawber and P. A. H. Wyatt J. Chem. Soc. **1960** 3589

D32 H. M. Dawson J. Chem. Soc. **1927** 1290

D33 J. S. Day and P. A. H. Wyatt J. Chem. Soc. **B** **1966** 343

D34 A. Delannoy, J. Hennion, J. C. Bavary and J. Nicole C.R. Acad. Sci. Ser. C **289** 401 (1979)

D35 H. G. Denham Z. Anorg. Allgem. Chem. **57** 378 (1903)

D36 H. G. Denham J. Chem. Soc. **93** 41 (1908)

D37 H. G. Denham and N. A. Marris Trans. Faraday Soc. **24** 510 (1928)

D38 T. O. Denney and C. B. Monk Trans. Faraday Soc. **47** 992 (1951)

D39 N. C. Deno J. Am. Chem. Soc. **74** 2039 (1952)

D40 N. C. Deno, H. E. Berkheimer, W. L. Evans and H. J. Peterson J. Am. Chem. Soc. **81** 2344 (1959)

D41 N. C. Deno, H. E. Berkheimer, W. L. Evans and H. J. Peterson J. Am. Chem. Soc. **81** 2344 (1959), recalculating data due to K. Singer and P. A. Vampler J. Chem. Soc. **1956** 3971 and T. A. Turney and G. A. Wright J. Chem. Soc. **1958** 2415

D42 N. C. Deno, J. J. Jaruzelski and A. Schriesheim J. Am. Chem. Soc. **77** 3044 (1955)

D43 N. C. Deno and N. H. Potter, J. Am. Chem. Soc. **89** 3550 (1967)

D44 N. C. Deno and R. W. Taft J. Am. Chem. Soc. **76** 244 (1954)

D45 B. Desire U.S. At. Energy Comm. 1970, NP-18284, Nucl. Sci. Abstr. **24** 36203 (1970)

D46 B. Desire, M. Hussonnais and R. Guillaumont, C.R. Acad. Sci., Ser. C **269** 448 (1969)

D47 D. Devèze C.R. Acad. Sci., Ser. C **263** 392 (1966)

D48 D. Devèze and P. M. Rumpf C.R. Acad. Sci. **258** 6135 (1964)

D49 H. DeVoe and G. B. Kistiakowsky J. Am. Chem. Soc. **83** 274 (1961)

D50 L. N. Devonshire and H. H. Rowley Inorg. Chem. **1** 680 (1962)

D51 N. Dhar and A. K. Datta Z. Elektrochem. **19** 407 (1913)

D52 H. Diebler and N. Sutin J. Phy. Chem. **68** 174 (1964)

D53 H. G. Dietrich and J. Johnson J. Am. Chem. Soc. **49** 1419 (1927)

D54 T. P. Dirske J. Electrochem. Soc. **101** 328 (1954)

D55 J. V. Dobson Chem. Ind. (London) **1970** 501

D56 J.V. Dobson and H. R. Thirsk Electrochim. Acta **16** 315 (1971)

D57 R. G. Downing and D. E. Pearson J. Am. Chem. Soc. **83** 1718 (1961)

D58 G. J. Doyle and N. Davidson J. Am. Chem. Soc. **71** 3491 (1949)

D59 C. Dragulescu, A. Nimara and L. Julean Chem. Anal. (Warsaw) **17** 631 (1972)

D60 C. Dragulescu, A. Nimara and L. Julean Rev. Roum. Chem. **17** 1181 (1972)

D61 N. S. Drozdov and V. P. Krylov Zhur. Fiz. Khim. **35** 2557 (1961)

D62 C. Drucker Trans. Faraday Soc. **33** 660 (1937)

D63 L. P. Ducret Ann. Chim. (Paris) [**12**] **6** 705 (1951)

D64 L. P. Ducret Ann. Chim. (Paris) [**12**] **6** 764 (1951)

D65 S. M. Dumpierrez and F. Brito Anales de Quim. **64** 115 (1968)

D66 J. F. Duncan and D. L. Kepert J. Chem. Soc. **1962** 205

D67 H. B. Dunford and W. D. Hewson J. Phys. Chem. 83 3307 (1979)

D68 H. S. Dunsmore, S. Hietanen and L. G. Sillén Acta Chem. Scand. 17 2644 (1963)

D69 H. S. Dunsmore and G. H. Nancollas J. Phys. Chem. 68 1579 (1964)

D70 J. Duplessis and R. Guillaumont Radiochem. Radioanal. Lett. 31 293 (1977)

D71 J. R. Durig, O. D. Bonner and W. H. Breazeale J. Phys. Chem. 69 3886 (1965)

D72 J. L. Dye, M. P. Faber and D. J. Karl J. Am. Chem. Soc. 82 314 (1960)

D73 D. Dyrssen and P. Lumme Acta Chem. Scand. 16 1785 (1962)

D74 D. Dyrssen and T. Sekine J. Inorg. Nucl. Chem. 26 981 (1964)

D75 D. Dyrssen and V. Tyrrell Acta Chem. Scand. 15 393, 1622 (1961)

E

E1 J. E. Earley, D. Fortnum, A. Wojcicki and J. O. Edwards J. Am. Chem. Soc. 81 1295 (1959)

E2 L. Ebert Naturwiss. 13 393 (1925)

E3 L. Ebert Monatsh. 80 788 (1949)

E4 J. T. Edward and I. C. Wang Canad. J. Chem. 40 399 (1962)

E5 J. T. Edward and I. C. Wang Canad. J. Chem. 43 2867 (1965)

E6 J. O. Edwards J. Am. Chem. Soc. 75 6151 (1953)

E7 O. W. Edwards, T. D. Farr, R. L. Dunn and J. D. Hatfield J. Chem. Eng. Data 18 24 (1973)

E8 A. M. Egorov and Z. K. Odinets Sb. Nauchn. Tr., Gos. Nauchn.-Issled. Inst. Tsvetn. Metal., 23 241 (1965)

E9 M. Ehrenfreund and J. L. Leibenguth Bull. Soc. Chim. France 1970 2498

E10 E. Eichler and S. Rabideau J. Am. Chem. Soc. 77 5501 (1955)

E11 M. Eigen and K. Kustin J. Am. Chem. Soc. 84 1355 (1962)

E12 H. Einaga J. Chem. Soc. Dalton Trans. 1977 912

E13 H. Einaga J. Chem. Soc. Dalton Trans. 1979 1917

E14 A. A. Elesin, I. A. Lebedev, E.M. Piskunov and G. N. Yakovlev Radiokhimiya 9 161 (1967)

E15 J. S. Elliot, R. F. Sharp and L. Lewis J. Phys. Chem. 62 686 (1958)

E16 A. J. Ellis Am. J. Sci. 257 287 (1959)

E17 A. J. Ellis J. Chem. Soc. 1959 3689

E18 A. J. Ellis J. Chem. Soc. 1963 4300

E19 A. J. Ellis and D. W. Anderson J. Chem. Soc. 1961 1765

E20 A. J. Ellis and D. W. Anderson J. Chem. Soc. 1961 4678

E21 A. J. Ellis and W. Giggenbach Geochim. Cosmochim. Acta 35 247 (1971)

E22 A. J. Ellis and R. M. Golding J. Chem. Soc. 1959 127

E23 A. J. Ellis and N. B. Milestone Geochim. Cosmochim. Acta 11 615 (1967)

E24 H. R. Ellison, J. O. Edwards and E. A. Healy J. Am. Chem. Soc. 84 1820 (1962)

E25 H. Ellison and A.E. Martell J. Inorg. Nucl. Chem. 26 1555 (1964)

E26 K. L. Elmore, J. D. Hatfield, R. L. Dunn and A. D. Jones J. Phys. Chem. 69 3520 (1965)

E27 K. Emerson and W. M. Graven J. Inorg. Nucl. Chem. 11 309 (1959)

E28 F. Ender, W. Teltschik and K. Schäfer Z Elektrochem. 61 775 (1957)

E29 A. Engelbrecht and B. M. Rode Monatsch. Chem. 103 1315 (1972)

E30 A. A. Ennan and V. A. Lapshin Zhur. Strukt. Khim. 14 21 (1973)

E31 A. G. Epprecht Helv. Chim. Acta 21 205 (1938)

E32 T. Erdey-Gruz, L. Majthenye and E. Kugler Acta Chim. Acad. Sci. Hung. 37 393 (1963)

E33 C. E. Evans and C. B. Monk Trans. Faraday Soc. 67 2652 (1971)

E34 M. G. Evans and N. Uri Trans. Faraday Soc. 45 224 (1949)

E35 D. H. Everett and D. A. Landsman Trans. Faraday Soc. 50 1221 (1954)

E36 A. J. Everett and G. J. Minkoff Trans. Faraday Soc. 49 410 (1953)

E37 D. H. Everett and W. F. K. Wynne-Jones Proc. Roy. Soc. (London) A169 190 (1938)

E38 D. H. Everett and W. F. K. Wynne-Jones Trans. Faraday Soc. 48 531 (1932)

F

F1 J. P. Fackler and I. D. Chawla Inorg. Chem. 3 1130 (1964)

F2 M. T. Falqui, G. Ponticelli and F. Sotgui Ann. Chim. (Rome) 56 464 (1966)

F3 L. Farkas and M. Lewin J. Am. Chem. Soc. 72 5766 (1950)

F4 H. N. Farrer and F. J. C. Rossotti J. Inorg. Nucl. Chem. 26 1959 (1964)

F5 J. Faucherre Bull. Soc. Chim. France 1954 128

F6 J. Faucherre Bull. Soc. Chim. France 1954 253

F7 C. Faurholt J. Chim. Phys. 21 400 (1924)

F8 A. Fava and G. Pajaro J. Am. Chem. Soc. 78 5203 (1956)

F9 P. Favier and R. Schaal Compt. Rend. 249 1231 (1959)

F10 E. H. Fawcett and S. F. Acree J. Res. Natl. Bur. Stand. 6 757 (1931)

F11 D. Feakins, W. A. Last and R. A. Shaw Chem. Ind. 1962 510

F12 D. Feakins, W. A. Last and R. A. Shaw J. Chem. Soc. 1964 4464

F13 L. R. Fedor and T. C. Bruice J. Am. Chem. Soc. 86 4117 (1964)

F14 V. A. Fedorov, T. N. Kalosh and N. R. Dergyazina Zhur. Neorg. Khim. 24 2317 (1979)

F15 F. Feher and F. Viel, Z. Anorg. Chem. 335 113 (1965)

F16 W. Feldman, Z. Anorg. Chem. 338 235 (1965)

F17 W. Feldman and E. Thilo Z. Anorg. Chem. 328 113 (1964)

F18 R. de G. Ferreira Anals. Acad. Brazil Cienscias 29 353 (1957)

F19 G. Ferroni Electrochim. Acta 21 283 (1976)

F20 G. Ferroni, J. Galea, G. Antonetti and R. Romanetti Bull. Soc. Chim. France 1974 2695

F21 A. Ferse Z. Physiol. Chem. (Leipzig) 229 51 (1965)

F22 Yu. Ya. Fialkov and Yu. Ya. Borovikov Vestn. Kievsr Politekhn. Inst., Ser. Khim. Mashinostroit. i. Tekhnol. 1965 66

F22a P. Fiordiponti, F. Rollo and F. Rodante Z. Phys. Chem. (Frankfurt) 88 149 (1974)

F23 R. Fischer and J. Byl Bull. Soc. Chim. France 1964 2920

F24 J. R. Fisher Diss. Abs. Int. B **30** 4942 (1970)
F25 J. R. Fisher and H. L. Barnes J. Phys. Chem. **76** 90 (1972)
F26 A. N. Fletcher J. Inorg. Nucl. Chem. **26** 955 (1964)
F27 E. P. Flint and L. S. Wells J. Res. Natl. Bur. Stand. **12** 752 (1934)
F28 I. E. Flis Russ. J. Applied Chem. (Engl. trans.) **31** 1183 (1958)
F29 I. E. Flis, K. P. Mishchenko and N. V. Pakhomova Zhur. Neorg. Khim. **3** 1772 (1958)
F30 I. E. Flis, K. P. Mishchenko and G. I. Pusenok Izv. Vysshikh Uchebn. Zavedenii Khim. i Khim. Tekhnol. **7** 764 (1964)
F31 I. E. Flis, K. P. Mishchenko and T. A. Tumanova Zhur. Neorg. Khim. **4** 277 (1959)
F32 P. Flood, T. J. Lewis and D. U. Richards J. Chem. Soc. **1963** 2446
F33 R. H. Flowers, R. J. Gillespie, J. V. Dubridge and C. Solomons J. Chem. Soc., **1958** 667
F34 S. Fontana and F. Brito Inorg. Chim. Acta **2** 179 (1968)
F35 J. S. Forrester and G. H. Ayres J. Phys. Chem. **63** 1979 (1959)
F36 W. Forsling, S. Hietanen and L. G. Sillén Acta Chem. Scand. **6** 901 (1952)
F37 D. H. Fortnum, C. J. Battaglia, S. R. Cohen and J. O. Edwards J. Am. Chem. Soc. **82** 778 (1960)
F38 F. Fouasson Ann. Chim. (Paris) [**12**] **3** 594 (1948)
F39 J. Francavilla and N. D. Chasteen, Inorg. Chem. **14** 2860 (1975)
F40 E. U. Franck Z. Physik. Chem. (Frankfurt) **8** 192 (1956)
F41 E. U. Franck, D. Hartmann and F. Hensel Disc. Faraday Soc. **39** 200 (1965) (but see J. V. Dobson, Chem. Ind. (London) **1970** 501)
F42 P. Franzosini, C. Sinistri and G. Airoldi Ricerca Sci. **30** 1707 (1960)
F43 R. T. M. Frazer J. Chem. Soc. **1965** 1747
F44 K. Fredenhagen and M. Wellman Z. Physik. Chem. **162** 458 (1932)
F45 V. Frei, J. Podlahová and J. Podlaha Collection Czech. Chem. Commun. **29** 2587 (1964)
F46 V. Frei and A. Ustianovicova Russ. J. Phys. Chem. (Engl. trans.) **37** 612 (1963)
F47 R. Frick and K. Meyring Z. Anorg. Allgem. Chem. **176** 325 (1928)
F48 C. R. Frink and M. Peech Inorg. Chem. **2** 473 (1963)
F49 U. K. Frolova, V. N. Kumok and V. V. Serebrennikov Izv. Vysshikh Uchebn. Zavedenii, Khim. i Khim. Technol., **9** 176 (1966); CA **65** 9816 (1966)
F50 M. Frydman, G. Nilsson, T. Rengemo and L. G. Sillén Acta Chem. Scand. **12** 878 (1958)
F51 C. R. Fuget Thesis, Pennsylvania State Univ., **1956**
F52 J. W. Fulton and D. J. Swinehart J. Am. Chem. Soc. **76** 864 (1954)
F53 S. C. Furman and J. T. Denison, unpubl., quoted by S.C. Furman and C. S. Garner, J. Am. Chem. Soc. **72** 1785 (1950)
F54 R. M. Fuoss and C. A. Kraus J. Am. Chem. Soc. **55** 476 (1933)
F55 A. Fürth Z. Elektrochem. **28** 57 (1922)
F56 G. Fuseya J. Am. Chem. Soc. **42** 368 (1920)

F57 W. S. Fyfe J. Chem. Soc. **1955** 1347

G

G1 H. Galal-Gorchev and W. Stumm J. Inorg. Nucl. Chem. **25** 567 (1963)

G2 J. Galea, G. Antonetti, G. Ferony and R. Romanetti Rev. Chim. Miner. **10** 475 (1973)

G3 J. Galea and J. Haladjian Rev. Chim. Miner. **7** 623 (1970)

G4 J. Galea, N. Sabiani, G. Antonetti and G. Perroni Ann. Chim. (Paris) **10** 155 (1975)

G5 J. M. Gallart Anales Real. Soc. Espan., Fiz. Quim. **31** 422 (1933)

G6 H. Gamsjaeger, K. Aeberhard and P. Schindler Helv. Chim. Acta **52** 2315 (1969)

G7 H. Gamsjaeger and P. Beutler J. Chem. Soc. Dalton Trans. **1979** 1415

G8 E. Sh. Ganelina and V. A. Borgoyakov Zhur. Neorg. Khim. **16** 596 (1971)

G9 L. A. Garmash and K. A. Barkov Deposited Doc. **1977**, Viniti 3433-77, 159

G9a A. B. Garrett and R. E. Heiks J. Am. Chem. Soc. **63** 562 (1941)

G10 A. B. Garrett and A. E. Hirschler J. Am. Chem. Soc. **60** 299 (1938)

G11 A. B. Garrett, O. Holmes and A. Laube J. Am. Chem. Soc. **62** 2024 (1940)

G12 A. B. Garrett and W. W. Howell J. Am. Chem. Soc. 61 1730 (1939)

G13 A. B. Garrett, S. Vellenga and C. M. Fontana J. Am. Chem. Soc. **61** 367 (1939)

G14 R. Gary, R. G. Bates and R. A. Robinson J. Phys. Chem. **68** 3806 (1964)

G15 V. Gaspar and M. Beck Magy. Kem. Foly. **86** 177 (1980)

G16 G. Gattow and M. Drager Z. Anorg. Chem. **349** 302 (1967)

G17 G. Gattow and B. Krebs Z. Anorg. Allgen. Chem. **323** 13 (1963)

G18 G. Gattow and J. Wortmann Z. Anorg. Chem. **345** 172 (1966)

G19 O. Gawron, S. Mahboob and J. Fernando J. Am. Chem. Soc. **86** 2283 (1964)

G20 K. H. Gayer and H. Leider J. Am. Chem. Soc. **77** 1448 (1955)

G21 K. H. Gayer and L. Wootner J. Am. Chem. Soc. **74** 1436 (1952)

G22 K. H. Gayer and L. Wootner J. Am. Chem. Soc. **78** 3944 (1956)

G23 K. H. Gayer and L. Wootner J. Phys. Chem. **61** 364 (1957)

G24 K. H. Gayer and O. T. Zajicek J. Inorg. Nucl. Chem. **26** 951 (1964)

G25 M. Gazikalovic, Z. Pavlovic and T. Markovic Arh. Tehnol. **5** 51 (1967); CA **70** 51292z

G26 R. W. Gelbach and G. B. King J. Am. Chem. Soc. **64** 1054 (1942)

G27 A. I. Gel'bshtein, R. P. Airapetova, G. G. Shcheglova and M. I. Temkin Zhur. Neorg. Khim. **9** 1502 (1964); A. I. Gel'bshtein, G. G. Shcheglova and M. I. Temkin Doklady Akad. Nauk S.S.S.R. **107** 108 (1965)

G28 A. I. Gel'bshtein, G. G. Shcheglova and M. I. Temkin Zhur. Neorg. Khim. **1** 282 (1956)

G29 A. I. Gel'bshtein, G. G. Shcheglova and M. I. Temkin Zhur. Neorg. Khim. **1** 506 (1956)

G30 R. E. George and E. M. Woolley J. Solution Chem. **1** 279 (1972)

G31 W. L. German, unpubl., quoted by H. T. S. Britton and R. A. Robinson J. Chem. Soc. **1931** 470

G32 T. K. Ghosh Current Sci. 40 541 (1970)
G33 A. K. Ghosh, J. C. Ghosh and B. Prasad J. Indian Chem. Soc. 57 1194 (1980)
G34 E. C. Gilbert J. Phys. Chem. 33 1235 (1929)
G35 R. J. Gillespie J. Chem. Soc. 1950 2493
G36 R. J. Gillespie J. Chem. Soc. 1950 2516
G37 R. J. Gillespie J. Chem. Soc. 1950 2537
G38 R. J. Gillespie, J. E. Peel and E. A. Robinson J. Am. Chem. Soc. 93 5083 (1971)
G39 R. J. Gillespie, E. A. Robinson and C. Solomons J. Chem. Soc. 1960 4320
G40 F. G. R. Gimblett and C. B. Monk Trans. Faraday Soc. 50 965 (1954)
G41 F. Giordani Gazz. Chim. Ital. 54 844 (1924)
G42 J. K. Gjaldback Z. Anorg. Allgem. Chem. 144 269 (1924)
G43 P. K. Glasoe and F. A. Long J. Phys. Chem. 64 188 (1960)
G44 V. A. Glebov, A. E. Klygin, J. D. Smirnova and N. S. Kolyada Zhur. Neorg. Khim. 17 3312 (1972)
G45 O. Glemser, W. Holznagel and S. I. Ali Z. Naturforsch. 20b 192 (1965)
G46 S. Gobom Acta Chem. Scand. A30 745 (1976)
G47 V. Gold and B. M. Lowe Proc. Chem. Soc. 1963 140
G48 V. Gold and B. M. Lowe J. Chem. Soc. A 1967 936
G49 V. Gold and C. H. Rochester J. Chem. Soc. 1964 1692, 1697, 1722, 1727
G50 M. Goldblatt and W. M. Jones J. Chem. Phys. 51 1881 (1969)
G51 P. Goldfinger and H. D. Graf von Schweinitz Z. Physik. Chem. B19 219 (1932)
G52 R. M. Golding J. Chem. Soc. 1960 3711
G53 S. Goldman, R. G. Bates and R. A. Robinson J. Solution Chem. 3 593 (1974)
G54 F. I. Golovin Trudy Molodykh Nauch. Robotnikov Akad. Nauk Kirgiz S.S.R. 1958 119; CA 55 4130 (1961)
G55 J. F. Goodman and P. Robson J. Chem. Soc. 1963 2871
G56 M. Gorman J. Am. Chem. Soc. 61 3342 (1939)
G57 Yu. G. Goroshchenko and R. V. Kuprina Zhur. Neorg. Khim. 22 1255 (1977)
G58 G. W. Goward Thesis, Princeton Univ., 1954
G59 R. W. Green and P. W. Alexander Austral. J. Chem. 18 651 (1965)
G60 S. A. Greenberg J. Am. Chem. Soc. 80 6508 (1958)
G61 S. A. Greenberg and E. W. Price J. Phys. Chem. 61 1539 (1957)
G62 E. Greenwald and D. W. Fong J. Phys. Chem. 73 650 (1969)
G63 A. D. Grieve, G. W. Gurd and O. Maass Canad. J. Res. 8 577 (1933)
G64 R. O. Griffith and A. McKeown Trans. Faraday Soc. 30 530 (1934)
G65 R. O. Griffith and A. McKeown Trans. Faraday Soc. 36 766 (1940)
G66 R. O. Griffith, A. McKeown and R. P. Taylor Trans. Faraday Soc. 36 752 (1940)
G67 A. K. Grzybowski J. Phys. Chem. 62 555 (1938)
G68 A. O. Gubeli and J. Ste-Marie Canadian J. Chem. 45 827 (1967)
G69 E. A. Guggenheim Trans. Faraday Soc. 62 2750 (1966)

G70 E. A. Guggenheim and T. D. Schindler J. Phys. Chem. 38 533 (1934)

G71 R. Guillaumont Bull. Soc. Chim. France 1965 135

G72 R. Guillaumont Compt. Rend. 260 1416 (1965)

G73 R. Guillaumont, Bull. Soc. Chem. France 1968 168

G74 R. Guillaumont, B. Desire and M. Galin Radiochem. Radioanal. Lett. 8 189 (1970)

G75 R. Guillaumont, C. Ferreira de Miranda and M. Crelin C.R. Acad. Sci. (Paris), Sect. C, 268 140 (1969)

G76 R. Guillaumont, J. C. Franck and R. Muxart Radiochem. Radioanal. Lett. 4 73 (1970)

G77 H. Guiter Ann. Chim. (Paris) [12] 2 72 (1947)

G78 H. Guiter Bull. Soc. Chim. France 1947 64

G79 H. Guiter Bull. Soc. Chim. France 1947 269

G80 O. E. Gulezian and J. H. Müller J. Am. Chem. Soc. 54 3141, 3151 (1932)

G81 R. L. Gustafson, C. Richard and A. E. Martell J. Am. Chem. Soc. 82 1526 (1960)

G82 R. Gut, J. Fluorine Chem. 15 163 (1980)

H

H1 J. Haas, N. Konopik, F. Mark and A. Neckel Monatsh. 95 1173 (1964)

H2 R. Haase, K. H. Dücker and H. A. Küppers Ber. Bunsenges. Physik. Chem. 69 97 (1965)

H3 G. Hägg Z. Anorg. Allgem. Chem. 155 21 (1926)

H4 H. Hagisawa Bull. Inst. Phys. Chem. Res. (Tokyo) 18 260 (1939)

H5 H. Hagisawa Bull. Inst. Phys. Chem. Res. (Tokyo) 18 368 (1939)

H6 H. Hagisawa Bull. Inst. Phys. Chem. Res. (Tokyo) 18 648 (1939)

H7 H. Hagisawa Bull. Inst. Phys. Chem. Res. (Tokyo) 19 1220 (1940)

H8 H. Hagisawa Bull. Inst. Phys. Chem. Res. (Tokyo) 20 251 (1941)

H9 H. Hagisawa Bull. Inst. Phys. Chem. Res. (Tokyo) 20 384 (1941)

H10 F. L. Hahn and R. Klockmann Z. Physik. Chem. A146 373 (1930)

H11 F. L. Hahn and R. Klockmann Z. Physik. Chem. A151 80 (1930)

H12 G. P. Haight, D. C. Richardson and N. H. Coburn Inorg. Chem. 3 1777 (1964)

H13 J. Hala, O. Navrátil and V. Nechuta J. Inorg. Nucl. Chem. 28 553 (1966)

H13a J. Haladjan, R. Sabbah and P. Bianco J. Chim. Phys. Physicochim. Biol. 65 1751 (1968); CA 70 51218c

H14 H. V. Halban and J. Brüll Helv. Chim. Acta 27 1719 (1944)

H15 M. Halmann, A. Lapidot and D. Samuel J. Chem. Soc. 1963 1299

H16 S. D. Hamann J. Phys. Chem. 67 2233 (1963)

H17 S. D. Hamann and W. Strauss Trans. Faraday Soc. 51 1684 (1955)

H18 W. J. Hamer J. Am. Chem. Soc. 56 860 (1934)

H19 S. A. Hamid, I. P. Alimarin and I. V. Puzdrenkova Vestn. Mosk. Univ., Ser. II, Khim. 20 71 (1965); CA 63 6376 (1965)

H20 L. P. Hammett and A. J. Deyrup J. Am. Chem. Soc. 54 2721 (1932)

H21 G. I. H. Hanania, D. H. Irvine, W. A. Eaton and P. George J. Phys. Chem. 71 2022 (1967)
H22 R. D. Hancock S. Afr. J. Chem. 32 49 (1979)
H23 L. D. Hansen, J. A. Partridge, R. M. Izatt and J. J. Christensen Inorg. Chem. 5 569 (1966)
H24 I. Hansson Acta Chem. Scand. 27 924 (1973)
H25 I. Hansson Acta Chem. Scand. 27 931 (1973)
H26 A. Hantzsch Ber. 32 3066 (1899)
H27 G. Harbottle J. Am. Chem. Soc. 73 4024 (1951)
H28 G. Harbottle and R. W. Dodson J. Am. Chem. Soc. 73 2442 (1951)
H29 M. K. Hargreaves and P. J. Richardson J. Chem. Soc. 1958 3111
H30 M. K. Hargreaves, R. A. Stevenson and J. Evans J. Chem. Soc. 1965 4582
H31 T. R. Harkins and H. Freiser J. Am. Chem. Soc. 77 1374 (1955)
H32 R. W. Harman J. Phys. Chem. 31 616 (1927)
H33 R. W. Harman and F. P. Worley Trans. Faraday Soc. 20 502 (1924)
H34 H. S. Harned and F. T. Bonner J. Am. Chem. Soc. 67 1026 (1945)
H35 H. S. Harned and H. R. Copson J. Am. Chem. Soc. 55 2206 (1933)
H36 H. S. Harned and R. Davis J. Am. Chem. Soc. 65 2030 (1943)
H37 H. S. Harned and J. G. Donelson J. Am. Chem. Soc. 59 1280 (1937)
H38 H. S. Harned and C. G. Geary J. Am. Chem. Soc. 59 2032 (1937)
H39 H. S. Harned and W. J. Hamer J. Am. Chem. Soc. 55 2194 (1933)
H40 H. S. Harned and W. J. Hamer J. Am. Chem. Soc. 55 4496 (1933)
H41 H. S. Harned and G. E. Mannweiler J. Am. Chem. Soc. 57 1873 (1935)
H42 H. S. Harned and B. B. Owen J. Am. Chem. Soc. 52 5079 (1930)
H43 H. S. Harned and R. A. Robinson Trans. Faraday Soc. 36 973 (1940)
H44 H. S. Harned and S. R. Scholes J. Am. Chem. Soc. 63 1706 (1941)
H45 W. E. Harris and I. M. Kolthoff J. Am. Chem. Soc. 69 446 (1947)
H46 W. H. Hartford Ind. Eng. Chem. 34 920 (1942)
H47 A. B. Hastings and J. Sendroy J. Biol. Chem. 65 445 (1925)
H48 E. M. Hattox and T. DeVries J. Am. Chem. Soc. 58 2126 (1936)
H49 J. A. Hearne and A. G. White J. Chem. Soc. 1957 2168
H50 K. H. Heckner and R. Lansberg J. Inorg. Nucl. Chem. 29 423 (1967)
H51 B. O. A. Hedström Arkiv. Kemi. 5 457 (1953)
H52 B. O. A. Hedström Arkiv. Kemi. 6 1 (1953)
H53 L. J. Heidt J. Phys. Chem. 46 624 (1942)
H54 E. Heilbronner and S. Weber Helv. Chim. Acta 32 1513 (1949)
H55 K. Heinzinger and R. E. Weston, J. Chem. Phys. 42 272 (1965)
H56 H. C. Helgeson J. Phys. Chem. 71 3121 (1967)
H57 R. P. Henry, P. C. H. Mitchell and J. E. Prue J. Chem. Soc., Dalton Trans. 1973 1156
H58 Y. Hentola Kemian Keskulüton Julkaisuja 13 No. 2 (1946)
H59 L. G. Hepler and Z. Z. Hugus J. Am. Chem. Soc. 74 6115 (1952)
H60 H. G. Hertz Ber. Bunsenges. Phys. Chem. 84 613, 622 (1980)
H61 A. Heydweiller Ann. Physik. 28 503 (1909)
H62 J. Heyrovsky Trans. Faraday Soc. 19 692 (1923)

H63 S. Hiètanen Acta Chem. Scand. 10 1531 (1956)

H64 S. Hiètanen and E. Högfeldt Chem. Ser. 10 41 (1976)

H65 S. Hiètanen and L. G. Sillén Acta Chem. Scand. 6 747 (1952)

H66 S. Hiètanen and L. G. Sillén Acta Chem. Scand. 13 533 (1959)

H67 S. Hiètanen and L. G. Sillén Acta Chem. Scand. 18 1015 (1964)

H68 S. Hiètanen and L. G. Sillén Acta Chem. Scand. 18 1018 (1964)

H69 S. Hiètanen and L. G. Sillén Acta Chem. Scand. 22 265 (1968)

H70 J. Hill and A. McAuley J. Chem. Soc. A 1968 2405

H71 J. C. Hindman The Transuranium Elements, Natl. Nuclear Energy Ser. IV-14B (1949)

H72 R. L. Hinman J. Org. Chem. 23 1587 (1958)

H73 R. L. Hinman and J. Lang J. Am. Chem. Soc. 86 3796 (1964)

H74 B. F. Hitch and R. E. Mesmer J. Solution Chem. 5 667 (1976)

H75 M. de Hlasko J. Chim. Phys. 20 167 (1923)

H76 W. H. Ho Proc. Natl. Sci. Council Part 1 (Taiwan), 10 175 (1977)

H76a W. H. Ho Diss. Abst. Int. B 35 3853 (1975)

H77 E. Högfeldt Acta Chem. Scand. 17 785 (1963)

H78 E. Högfeldt and J. Bigeleisen J. Am. Chem. Soc. 82 15 (1960)

H79 B. Holmberg Z. Physik. Chem. 70 157 (1910)

H80 L. P. Holmes, D. L. Cole and E. M. Eyring J. Phys. Chem. 72 301 (1968)

H81 G. Holst Svensk. Kem. Tidskr. 52 258 (1940)

H82 G. Holst Svensk. Papperstidn. 48 23 (1945)

H83 W. B. Holzapfel J. Chem. Phys. 50 4424 (1969)

H84 W. Holzapfel and E. U. Franck Ber. Bunsenges. Phys. Chem. 70 1105 (1966)

H85 C. C. Hong and W. H. Rapson Canad. J. Chem. 46 2053 (1968)

H86 G. C. Hood, A. C. Jones and C. A. Reilly J. Phys. Chem. 63 101 (1959)

H87 G. C. Hood, O. Redlich and C. A. Reilly J. Chem. Phys. 22 2067 (1954)

H88 G. C. Hood and C. A. Reilly J. Chem. Phys. 32 127 (1960)

H89 H. P. Hopkins and C. A. Wulff J. Phys. Chem. 69 1980 (1965)

H90 R. A. Horne, R. A. Courant and G. R. Frysinger J. Chem. Soc. 1964 1515

H91 P. B. Hostetler Am. J. Sci. 261 238 (1963)

H92 J. R. Howard, V. S. K. Nair and G. H. Nancollas Trans. Faraday Soc. 54 1034 (1958)

H93 J. Höye Kgl. Norske Videnskab. Selskabs Forhandl. 14 1 (1942)

H94 K. H. Hsu and S. H. Ho K'o Hsüeh T'ung Pao 1957 433; CA 55 21953 (1961)

H95 J. Hudis and M. Wolfsberg, unpubl., quoted by J. Hudis and R. W. Dodson J. Am. Chem. Soc. 78 911 (1956)

H96 R. Hugel Bull. Soc. Chim. France 1964 1462

H97 R. Hugel Bull. Soc. Chim. France 1965 968

H98 W. S. Hughes J. Chem. Soc. 1928 491

H99 M. N. Hughes and G. Stedman J. Chem. Soc. 1963 1239

H100 A. Huss and C. A. Eckert J. Phys. Chem. 81 2268 (1977)

H101 H. Hussonnois, S. Hubert, L. Aubin, R. Guillaumont and G. Boussieres Radiochem. Radioanal. Lett. 10 231 (1972)

H102 H. Hussonnois, S. Hubert, L. Brillard and R. Guillaumont Radiochem. Radioanal. Lett. 15 47 (1973)

H103 H. H. Hyman, M. Kilpatrick and J. J. Katz J. Am. Chem. Soc. **79** 3668 (1957)

I

I1 A. Indelli and G. Mantovani Trans. Faraday Soc. **61** 909 (1965)
I2 J. W. Ingram and J. Morrison J. Chem. Soc. **1933** 1200
I3 N. Ingri Acta Chem. Scand. **13** 758 (1959)
I4 N. Ingri Acta Chem. Scand. **16** 439 (1962)
I5 N. Ingri Acta Chem. Scand. **17** 573 (1963)
I6 N. Ingri Acta Chem. Scand. **17** 581 (1963)
I7 N. Ingri Acta Chem. Scand. **17** 597 (1963)
I8 N. Ingri and F. Brito Acta Chem. Scand. **13** 1971 (1959)
I9 N. Ingri, G. Lagerström, M. Frydman and L. G. Sillén Acta Chem. Scand. **11** 1034 (1957)
I10 N. Ingri and G. Schorsch Acta Chem. Scand. **17** 590 (1963)
I11 L. N. Intshirvali, I. V. Kolozov and G. M. Varshal Zhur. Neorg. Khim. **20** 8 (1975)
I12 R. R. Irani J. Phys. Chem. **65** 1463 (1961)
I13 R. R. Irani and C. F. Callis J. Phys. Chem. **65** 934 (1961)
I14 R. R. Irani and T. A. Taulli J. Inorg. Nucl. Chem. **28** 1011 (1966)
I15 M. Isaks and H. H. Jaffé J. Am. Chem. Soc. **86** 2209 (1964)
I16 H. Isikawa and I. Aoki Bull. Inst. Phy., Chem. Research (Tokyo) **19** 136 (1940)
I17 Y. J. Israeli Bull. Soc. Chim. France **1965** 193
I18 I. M. Issa and S. A. Awad J. Phys. Chem. **58** 948 (1954)
I19 R. M. Issa, N. A. Ibrahim and B. A. Abd-El-Nabey Egypt. J. Chem. **16** 263 (1973)
I20 T. Ito and N. Yui Sci. Rpts. Tohoku Univ. **37** 19 (1953); CA **48** 6791 (1954)
I21 T. Ito and N. Yui Sci. Rpts. Tohoku Univ. **37** 185 (1953); CA **48** 5613 (1954)
I22 A. A. Ivakin, S. V. Vorob'eva and E. M. Gertman Zhur. Neorg. Khim. **24** 36 (1979)
I23 A. A. Ivakin, S. V. Vorob'eva, E. M. Gertman and E. M. Voranova Zhur. Neorg. Khim. **21** 442 (1976)
I24 A. A. Ivakin and E. M. Voronova, Zhur. Neorg. Khim. **18** 885 (1973)
I25 A. N. Ivanov and S. N. Aleshin Doklady Mokov Sel'skokhoz Akad. Nauch. Konfer **22** 386 (1956)
I26 G. F. Ivanova, N. I. Levkina, L. A. Nestorova, A. P. Zhidkova and U. Khodakovskii, Geokhimiya **2** 234 (1975)
I27 M. F. Ivanova and M. B. Neiman Doklady Akad. Nauk S.S.S.R. **60** 1005 (1948)
I28 B. N. Ivanov-Emin, L. A. Nisel'son and L. E. Larionova Russ. J. Inorg. Chem. (Engl. transl.) **7** 266 (1962)
I29 R. M. Izatt, J. J. Christensen and C. H. Bartholomew J. Chem. Soc. A **1969** 45

I30 R. M. Izatt, J. J. Christensen, R. T. Pack and R. Bench *Inorg. Chem.* **1** 828 (1962)

I31 R. M. Izatt, D. Eatough and J. J. Christensen *J. Chem. Soc.* **A** **1967** 1301

J

J1 H. H. Jaffé and R. W. Gardner *J. Am. Chem. Soc.* **80** 319 (1958)

J1a D. V. S. Jain and C. M. Jain *J. Chem. Soc.* **A** **1967** 1541

J2 G. Jander and F. Kienbaum *Z. Anorg. Allgem. Chem.* **316** 63 (1962)

J3 K. Jellinek *Z. Physik. Chem.* **76** 257 (1911)

J3a K. Jellinek and J. Czerwinski *Z. Physik. Chem.* **102** 438 (1922)

J4 M. B. Jensen *Acta Chem. Scand.* **12** 1657 (1958)

J5 B. A. Jillot and R. J. P. Williams *J. Chem. Soc.* **1958** 462

J6 P. Job *Compt. Rend.* **186** 1546 (1928)

J7 J. K. Johannesson *J. Chem. Soc.* **1959** 2998

J8 A. Johansson and E. Wänninen *Talanta* **10** 769 (1963)

J9 C. E. Johnson *J. Am. Chem. Soc.* **74** 959 (1952)

J10 G. K. Johnson and J. E. Bauman *Inorg. Chem.* **17** 2774 (1978)

J11 C. D. Johnson, A. R. Katritzky and S. A. Shapiro *J. Am. Chem. Soc.* **91** 6654 (1969)

J12 H. L. Johnston, F. Cüta and A. B. Garrett *J. Am. Chem. Soc.* **55** 2311 (1933)

J13 H. L. Johnston and H. L. Leland *J. Am. Chem. Soc.* **60** 1439 (1938)

J14 H. F. Johnstone and P. W. Leppla *J. Am. Chem. Soc.* **56** 2233 (1934)

J15 W. L. Jolly *J. Am. Chem. Soc.* **74** 6199 (1952)

J16 J. Jordan and G. J. Ewing *Inorg. Chem.* **1** 587 (1962)

J16a C. K. Jorgensen *Acta Chem. Scand.* **10** 502 (1956)

J17 E. Jorgensen and J. Bjerrum *Acta Chem. Scand.* **12** 1047 (1958)

J18 M. J. Jorgenson and D. R. Hartler *J. Am. Chem. Soc.* **85** 878 (1963)

J19 J. Jortner and G. Stein *Bull. Research Council Israel* **6A** 239 (1957)

J20 H. F. Joseph and H. B. Oakley *J. Chem. Soc.* **127** 2813 (1925)

J21 M. L. Josien and G. Sourisseau *Bull. Soc. Chim. France* **17** 255 (1950)

J22 M. Jowett and H. Millet *J. Am. Chem. Soc.* **51** 1004 (1929)

J23 M. Jowett and H. I. Price *Trans. Faraday Soc.* **28** 668 (1932)

J24 R. A. Joyner *Z. Anorg. Allgem. Chem.* **77** 103 (1912)

K

K0 Kabir-ul-Din *Z. Phys. Chem. (Frankfurt)* **88** 316 (1974)

K1 H. C. Kaehler, S. Mateo and F. Brito *An. Quim.* **69** 1273 (1973)

K2 H. C. Kaehler, S. Mateo and F. Brito *An. Quim.* **71** 688 (1975)

K3 H. Kakihana, T. Amaya and M. Maeda *Bull. Chem. Soc. Japan* **43** 3155 (1970)

K4 H. Kakihana and M. Maeda *Bull. Chem. Soc. Japan* **42** 1458 (1969)

K5 H. Kakihana and M. Maeda *Bull. Chem. Soc. Japan* **43** 109 (1970)

K6 H. Kakihana and L. G. Sillén *Acta Chem. Scand.* **10** 985 (1956)

K7 W. Kangro Z. Physik. Chem. (Frankfurt) 32 273 (1962)

K8 J. J. Kankara Anal. Chem. 44 2376 (1972)

K9 C. W. Kanolt J. Am. Chem. Soc. 29 1402 (1907)

K10 D. Kantzer Compt. Rend. 220 661 (1945)

K11 V. A. Kargin Z. Anorg. Allgem. Chem. 183 77 (1929)

K12 J. Kasper Thesis, Johns Hopkins Univ., Baltimore, 1941

K13 Y. Kauko Ann. Acad. Sci. Fennicae 39A No. 3 (1934); 41A No. 9 (1935)

K14 Y. Kauko and A. Airola Z. Physik. Chem. A179 307 (1937)

K15 Y. Kauko and J. Carlberg Z. Physik. Chel. A173 141 (1935)

K16 Y. Kauko and H. Elo Z. Physik. Chem. A184 211 (1939)

K17 Y. Kauko and V. Mantere Z. Physik. Chem. A176 187 (1936)

K18 T. Kawai, H. Otsuka and M. Maeda Bull. Chem. Soc. Japan 43 109 (1970)

K19 C. Th. Kawassiades, G. E. Manoussakis, O. Ch. Papavasillon and J. A. Tossidis Chim. Chron. A 33 4 (1968)

K20 C. T. Kawassiades, G. E. Manoussakis and J. A. Tossidis J. Inorg. Nucl. Chem. 29 401 (1967)

K21 C. M. Kelley and H. V. Tartar J. Am. Chem. Soc. 78 5752 (1956)

K22 B. Kennedy, Diss. Abs. Int. B 30 1544 (1969)

K23 J. Kenttämaa Ann. Acad. Sci. Fennicae 67A No. 2 (1955)

K24 J. Kenttämaa Suomen Kem. 30B 9 (1957)

K25 J. Kenttämaa Suomen Kem. 32B 55 (1959)

K26 M. Kerker J. Am. Chem. Soc. 79 3664 (1957)

K27 M. Kiese and A. B. Hastings J. Am. Chem. Soc. 61 1291 (1939)

K28 M. Kiese and A. B. Hastings J. Biol. Chem. 132 267 (1940)

K29 G. Kilde Z. Anorg. Allgem. Chem. 218 113 (1934)

K30 M. Kilpatrick and L. Pokras J. Electrochem. Soc. 100 85 (1953)

K31 M. Kilpatrick and L. Pokras J. Electrochem. Soc. 101 39 (1954)

K32 S. Kilpi, A. Nurmi and Y. Rinne Suomen Kem. 24B 31 (1951)

K33 J. King and N. Davidson J. Am. Chem. Soc. 80 1542 (1958)

K34 E. J. King and G. W. King J. Am. Chem. Soc. 74 1212 (1952)

K35 R. W. Kingerley and V. K. LaMer J. Am. Chem. Soc. 63 3256 (1941)

K36 B. J. Kirkbridge and P. A. H. Wyatt Trans. Faraday Soc. 54 483 (1958)

K37 E. Klarman Z. Anorg. Allgem. Chem. 132 289 (1924)

K38 A. Klemenc and E. Hayek Monatsh. 54 407 (1929)

K39 A. Klemenc and M. Herzog Monatsh. 47 405 (1926)

K40 R. Klement, G. Biberacher and V. Hille Z. Anorg. Allgem. Chem. 289 80 (1957)

K41 I. M. Klotz, Thesis, Univ. of Chicago, 1940

K42 A. E. Klygin, I. D. Smirnova and D. M. Zavrazhnova, Zhur. Neorg. Khim 15 294 (1970)

K43 N. F. Knapp, Diss. Abs. Int. B 33 5197 (1973)

K44 R. J. Knight and R. N. Sylva, J. Inorg. Nucl. Chem. 37 779 (1975)

K45 J. Knox Z. Elektrochem. 12 477 (1906)

K46 J. R. Kolczynski, E. M. Roth and E. S. Shanley J. Am. Chem. Soc. 79 531 (1957)

K47 G. R. Kolonin and S. A. Stepoachikova Issled. po Eksperim. Mineral. Novosibirsk 1978 170

K48 I. V. Kolosov, L. N. Intchinveli and G. M. Varshal Zhur. Neorg. Khim. 20 2121 (1975)

K48a G. E. Kolski and D. W. Margerum Inorg. Chem. 7 2239 (1968)

K49 I. M. Kolthoff Chem. Weekblad 14 1016 (1917)

K50 I. M. Kolthoff Chem. Weekblad 16 1154 (1919)

K51 I. M. Kolthoff Pharm. Weekblad 57 474 (1920)

K52 I. M. Kolthoff Pharm. Weekblad 57 514 (1920)

K53 I. M. Kolthoff Z. Anorg. Allgem. Chem. 110 143 (1920)

K54 I. M. Kolthoff Rec. Trav. Chim. 42 969 (1923)

K55 I. M. Kolthoff Rec. Trav. Chim. 42 973 (1923)

K56 I. M. Kolthoff Rec. Trav. Chim. 43 216 (1924)

K57 I. M. Kolthoff Rec. Trav. Chim. 46 350 (1927)

K58 I. M. Kolthoff and M. Bosch Rec. Trav. Chim. 46 180 (1927)

K59 I. M. Kolthoff and M. Bosch Rec. Trav. Chim. 47 819 (1928)

K60 I. M. Kolthoff and M. Bosch Rec. Trav. Chim. 47 826 (1928)

K61 I. M. Kolthoff and M. K. Chantooni J. Am. Chem. Soc. 90 5961 (1968)

K62 I. M. Kolthoff and T. Kameda J. Am. Chem. Soc. 53 832 (1931)

K63 I. M. Kolthoff and W. J. Tomsicek J. Phys. Chem. 39 955 (1935)

K64 N. P. Komar Uchenye Zapiski Khar'kov. Univ., 133 Trudy Khim. Fak. Kh. Gh. 19 189 (1963)

K65 N. P. Komar, V. A. Naumenko and T. A. Kerpova Zhur. Fiz. Khim. 48 1613 (1974)

K66 N. P. Komar and Z. A. Tretyak Zhur. Analit. Khim. 10 236 (1955)

K67 N. P. Komar and T. V. Vasil'eva Deposited Doc. 1976, VINITI 2914-36

K68 N. P. Komar S. I. Vouk and V. N. Podnus Zhur. Fiz. Khim 51 1312 (1977)

K69 L. N. Komissarova, N. M. Prutkova and G. Ya. Pushkina Zhur. Neorg. Khim. 16 1798 (1971)

K70 A. Komura, M. Hayashi and H. Imanaga Bull. Chem. Soc. Japan 50 2927 (1977)

K71 T. Ya. Konchkova Zhur. Prikl. Khim. (Leningrad) 48 1649 (1975)

K72 N. Konopik and O. Leberl Monatsh. 80 655 (1949)

K72a N. Konopik and O. Leberl Monatsh. 80 781 (1949)

K73 I. M. Korenman Zhur. Obshchei Khim. 21 1961 (1951)

K74 Yu. S. Korotkin Radiokhimiya 17 547 (1955)

K75 I. A. Korshunov and E. F. Khrul'kova Zhur. Obshchei Khim. 19 2045 (1949)

K76 P. Kosonen Ann. Univ. Turku., Ser. Al 1970 No. 143

K77 A. Kossiakoff and D. Harker J. Am. Chem. Soc. 60 2047 (1938)

K78 E. Koubek, M. L. Haggett, C. J. Battaglia, K. M. Ibne-Rasa, H. Y. Pyun and J. O. Edwards J. Am. Chem. Soc. 85 2263 (1963)

K79 J. Kragten and L. G. Decnop-Weever Talanta 25 147 (1978)

K80 J. Kragten and L. G. Decnop-Weever Talanta 26 1105 (1979)

K80a J. Kragten and L. G. Decnop-Weever Talanta 27 1047 (1980)

K81 K. A. Kraus and J. R. Dam U.S. Atomic Energy Commission Reports CN-2831 (1946); CL-P-432 (1945); CL-P-449 (1945)

K82 K. A. Kraus and J. R. Dam The Transuranium Elements, Natl. Nuclear Energy Ser. IV-14B (1949)

K83 K. A. Kraus and R. W. Holmberg J. Phys. Chem. 58 325 (1954)

K84 K. A. Kraus and F. Nelson U.S. Atomic Energy Commission Report CNL-19 (1948)

K85 K. A. Kraus and F. Nelson J. Am. Chem. Soc. 72 3901 (1950)

K86 K. A. Kraus and F. Nelson J. Am. Chem. Soc. 77 3721 (1955)

K87 C. A. Kraus and H. C. Parker J. Am. Chem. Soc. 44 2429 (1922)

K88 A. Krawetz Thesis, University of Chicago, 1955

K89 H. Krentzien and F. Brito Ion (Madrid) 30 (342) 14 (1970)

K90 A. F. Kreshkov, V. A. Drozdov and N. A. Kalchina, Zhur. Fiz. Khim., 40 2150 (1966)

K91 M. E. Krevinskaya, V. D. Nikol'skii, B. G. Pozharskii and E. E. Zastenker Radiokhimiya 1 548 (1959); CA 54 15046 (1960)

K92 G. Krüger and E. Thilo Z. Physik. Chem. A209 190 (1958)

K93 G. Krüger and E. Thilo Z. Anorg. Allgem. Chem. 308 342 (1961)

K94 E. G. Krunchak, A. G. Rodichev, Ya.S. Khvorostin, B. S. Krumgal'z, V. G. Krunchuk and Yu. I. Yusova, Zhur. Neorg. Khim 18 2859 (1973)

K95 P. A. Kryukov and L. I. Starostina Izv. Sib. Otd. Akad. Nauk. S.S.S.R., Ser. Khim. Nauk. 1978 84

K96 P.A. Kryukov, L. I. Starostina, S. Ya. Tarasenko, L. A. Pavlyuk, B. S. Smolyakova and E. G. Larionov Mezhdunan Geokhim. Kongr. [Dokl.] 1 1971, 2 186

K97 P. A. Kryukov, L. I. Starostina, S. Ya. Tarasenko and M. R. Primanchuk Geokhimiya 7 1003 (1974)

K98 Z. Ksandr and M. Hejtmanek Shornik I. Celostatni Pracovni Konf. Anal. Chemiku 1952 42; CA 50 3150 (1956)

K99 H. Kubli Helv. Chim. Acta 29 1962 (1946)

K100 H. Kubota Thesis, University of Wisconsin, 1956

K101 I. N. Kugelmass Biochem. J. 23 587 (1930)

K102 F. Ya. Kul'ba, E. A. Kopylov, Yu. B. Yakovlev and E. G. Kolerova Zhur. Neorg. Khim. 17 2604 (1972)

K103 F. Ya. Kul'ba, V. G. Ushakova, Yu. B. Yakovlev and I. Vitkavskaite Vopr. Khimii Rastvorow Elektrolitov 1977 52

K104 F. Ya. Kul'ba, Yu. B. Yakovlev and E. A. Kopylov Zhur. Neorg. Khim. 15 2112 (1970)

K105 F. Ya. Kul'ba, Yu. B. Yakovlev and D. A. Zenchenko, Zhur. Neorg. Khim. 19 923 (1974)

K106 F. Ya. Kul'ba, Yu. B. Yakovlev and D. A. Zenchenko, Zhur. Neorg. Khim. 20 1781 (1975)

K107 F. Ya. Kul'ba, D. A. Zenchenko and Yu. B. Yakovlev, Zhur. Neorg. Khim. 20 2645

K108 C. Kullgren Z. Physik. Chem. 85 466 (1913)

K109 F. Kunschert Z. Anorg. Allgem. Chem. 41 337 (1904)

K110 F. W. Küster and E. Heberlein Z. Anorg. Allgem. Chem. 43 53 (1905)

K111 J. R. Kyrki Suomen Kem. B38 203 (1965)

L

L1 S. Lacroix Ann. Chem. (Paris) **4** 5 (1949)

L2 G. Lagerström Acta Chem. Scand. **13** 722 (1959)

L3 L. H. J. Lajunen and S. Parhi Finn. Chem. Lett. **1979** 143

L4 P. E. Lake and J. M. Goodings Canad. J. Chem. **36** 1089 (1958)

L5 A. B. Lamb and G. R. Fonda J. Am. Chem. Soc. **43** 1154 (1921)

L6 A. B. Lamb and A. G. Jacques J. Am. Chem. Soc. **60** 1215 (1938)

L7 S. M. Lambert and J. I. Watters J. Am. Chem. Soc. **79** 4262 (1957)

L8 O. E. Lanford and S. J. Kiehl J. Phys. Chem. **45** 300 (1941)

L9 O. E. Lanford and S. J. Kiehl J. Am. Chem. Soc. **64** 291 (1942)

L10 E. Lanza and G. Carpeni Electrochim. Acta **13** 519 (1968)

L11 S. Lasztity Radiochem. Radioanal. Lett. **29** 215 (1977)

L12 W. M. Latimer Oxidation Potentials, 2nd edn., Prentice-Hall Inc., New York, **1952** 86

L13 W. M. Latimer and H. W. Zimmermann J. Am. Chem. Soc. **61** 1550 (1939)

L14 S. H. Laurie, J. M. Williams and C. J. Nyman J. Phys. Chem. **68** 113 (1965)

L15 R. W. Lawrence J. Am. Chem. Soc. **56** 776 (1934)

L16 V. I. Lazarev and Yu. V. Moiseev Zh. Fiz. Khim. **39** 445 (1965)

L17 D. G. Lee and R. Stewart Canad. J. Chem. **42** 486 (1964)

L18 D. G. Lee and R. Stewart J. Am. Chem. Soc. **86** 3051 (1964)

L19 J. Lefebvre J. Chim. Phys. **54** 567 (1957)

L20 J. Lefebvre J. Chim. Phys. **55** 227 (1958)

L21 J. Lefebvre and H. Maria Compt. Rend. **256** 3121 (1963)

L22 M. Leist Z. Physik. Chem. (Leipzig) **205** 16 (1955)

L23 D. Leonesi and G. Piantoni Ann. Chim. (Rome) **55** 668 (1965)

L24 D. L. Leussing and I. M. Kolthoff J. Am. Chem. Soc. **75** 2476 (1953)

L25 G. R. Levi and R. Curli Ricerca Sci. **23** 1798 (1953)

L26 R. S. Levin, V. V. Sergeeva and A. T. Kolyshev Izv. Sib. Otd. Akad. Nauk S.S.S.R., Ser. Khim. Nauk, **1975** 113

L27 D. Levine and I. R. Wilson Inorg. Chem. **7** 818 (1968)

L28 S. A. Levison and R. A. Marcus J. Phys. Chem. **72** 358 (1968)

L29 M. G. Levy Gazz. Chim. Ital. **31** **II** 1 (1901)

L30 D. Lewis Acta Chem. Scand. **17** 1891 (1963)

L31 G. N. Lewis, T. B. Brighton and R. L. Sebastian J. Am. Chem. Soc. **39** 2245 (1917)

L32 G. N. Lewis and P. W. Schutz J. Am. Chem. Soc. **56** 1913 (1934)

L33 N. C. C. Li and Y. T. Lo J. Am. Chem. Soc. **63** 397 (1941)

L34 J. C. M. Li and D. M. Ritter J. Am. Chem. Soc. **75** 5823 (1963)

L35 J. C. M. Li and D. M. Ritter J. Am. Chem. Soc. **75** 5828 (1963)

L36 A. Liberti, V. Chiantella and F. Corigliano J. Inorg. Nucl. Chem. **25** 415 (1963)

L37 H. A. Liebhafsky J. Am. Chem. Soc. **61** 3513 (1939)

L38 H. A. Liebhafsky and B. Makower J. Phys. Chem. **37** 1037 (1933)

L39 M. H. Lietzke and R. W. Stoughton J. Phys. Chem. **67** 652 (1963)

L40 M. H. Lietzke, R. W. Stoughton and T. F. Young J. Phys. Chem. 65 2247 (1961)

L40a J. Lindner Monatsh. Chem. 33 613 (1912)

L40b W. T. Lindsay J. Phys. Chem. 66 1341 (1962)

L40c F. Lindstrand Svensk. Kem. Tidskr. 56 251 (1944)

L41 F. Lindstrand Svensk. Kem. Tidskr. 56 282 (1944)

L42 R. E. Lindstrom Diss. Abs. Int. B 28 4091 (1968)

L43 R. Lindstrom and H. E. Worth J. Phys. Chem. 73 218 (1969)

L44 J. J. Lingane and L. W. Niedrach J. Am. Chem. Soc. 70 4115 (1948)

L45 H. G. Linge and A. L. Jones Australian J. Chem. 21 1445, (2189) (1968)

L46 E. D. Linov and P. A. Kryukov Izv. Sib. Otd. Akad. Nauk S.S.S.R., Ser. Khim. Nauk (4) 10 (1972)

L47 M. W. Lister Canad. J. Chem. 30 879 (1952)

L48 M. W. Lister and Y. Yoshino Canad. J. Chem. 38 2342 (1960)

L49 V. M. Litvinchuk and K. N. Mikhalevich Ukr. Khim. Zhur. 25 563 (1959)

L50 S. Ljunggren and O. Lamm Acta Chem. Scand. 12 1834 (1958)

L51 V. I. Lobov, F. Ya. Kul'ba and V. I. Mironov Zhur. Neorg. Khim. 12 334 (1967)

L52 N. Löfman Z. Anorg. Allgem. Chem. 107 241 (1919)

L53 R. Lorenz and A. Böhl Z. Physik. Chem. 66 733 (1909)

L54 M. Lourijsen-Teyssèdre Bull. Soc. Chim. France 1955 1111

L55 M. Lourijsen-Teyssèdre Bull. Soc. Chim. France 1955 1118

L56 M. Lourijsen-Teyssèdre Bull. Soc. Chim. France 1955 1196

L57 H. L. Loy and D. M. Himmelblau J. Phys. Chem. 65 264 (1961)

L58 C. Luca and O. Enea Rev. Roum. Chim. 13 721 (1968)

L59 J. W. H. Lugg J. Am. Chem. Soc. 53 1 (1931)

L60 J. W. H. Lugg Trans. Faraday Soc. 27 297 (1931)

L61 O. Lukhari Suomen Kemistelehti B 43 347 (1974)

L62 P. Lumme, P. Lahermo and J. Tummavuori Acta Chem. Scand. 19 2175 (1965)

L63 P. Lumme and J. Tummavuori Acta Chem. Scand. 19 617 (1965)

L64 H. Lundén J. Chim. Phys. 5 574 (1907)

L65 R. Luther Z. Electrochem. 13 294 (1907)

L66 B. Lutz and H. Wendt Ber. Bunsenges. Phys. Chem. 74 372 (1970)

M

M1 A. H. J. Maas, A. N. P. Van Heijst and B. F. Visser Clin. Chim. Acta 33 325 (1971)

M2 D. H. Macartney and A. McAuley Inorg. Chem. 18 2891 (1979)

M3 J. C. McCoubrey Trans. Faraday Soc. 51 743 (1955)

M4 D. H. McDaniel and L. H. Steinert J. Am. Chem. Soc. 88 4826 (1966)

M5 D. D. Macdonald, P. Butler and D. Owen Canad. J. Chem. 51 2590 (1973)

M6 A. O. McDougall and F. A. Long J. Phys. Chem. 66 429 (1962)

M7 L. A. McDowell and H. L. Johnston J. Am. Chem. Soc. 59 2009 (1936)

M8 J. D. McGilvery and J. P. Crowther Canad. J. Chem. 32 174 (1954)

M9 D. A. MacInnes and D. Belcher J. Am. Chem. Soc. 55 2630 (1933)

M10 D. A. MacInnes and D. Belcher J. Am. Chem. Soc. 57 1683 (1935)

M11 H. A. C. McKay Trans. Faraday Soc. 52 1568 (1956)
M12 H. A. C. McKay and J. L. Woodhead J. Chem. Soc. 1964 717
M13 W. W. McNabb, J. F. Hazel and R. A. Baxter J. Inorg. Nucl. Chem. 30 1585 (1968)
M14 F. M. Mader J. Am. Chem. Soc., 80 2634 (1958)
M15 M. Maeda, T. Amaya and H. Kakihana Z. Naturforsch., B, 32B 1493 (1971)
M16 M. Maeda and H. Kakihana Bull. Chem. Soc. Japan 43 1097 (1970)
M17 M. Maeda, Y. Sunaoka and H. Kakihana J. Inorg. Nucl. Chem. 40 509 (1978)
M18 A. Maffei Gazz. Chim. Ital. 64 149 (1934)
M19 F. Maggio, V. Romano and L. Pellerito Ann. Chim. (Rome) 57 191 (1967)
M20 S. Mahapatra and R. S. Subrahmanya Proc. Indian Acad. Sci., Sect. A 65 283 (1967)
M21 O. Makitie and V. Konttinen Acta Chem. Scand. 23 1459 (1964)
M22 O. Makitie and M. L. Savolainen Suomen Kemistehleti 41B 242 (1968)
M23 I. N. Makslmova and V. F. Yushkevich Elektrokhimiya 2 577 (1966)
M24 H. C. Malhotra and S. K. Thereja Indian J. Chem. 14A 223 (1976)
M25 J. W. Malin and R. C. Koch Inorg. Chem. 17 752 (1978)
M26 L. J. Malone and R. W. Parry Inorg. Chem. 6 217 (1967)
M27 H. C. Mandell and G. Barth-Wehrenalp J. Inorg. Nucl. Chem. 12 90 (1959)
M28 G. G. Manov, N. J. DeLollis and S. F. Acree J. Research Natl. Bur. Standards 33 287 (1944)
M29 Y. Marcus Acta Chem. Scand. 11 690 (1957)
M30 Y. Marcus J. Chem. Soc. Faraday Trans.I, 75 1715 (1979)
M31 B. Marin and T. Kikindai C.R. Acad. Sci., Paris, Ser. C 268 1 (1969)
M32 W. Mark Acta Chem. Scand. A 31 157 (1977)
M33 G. Maronny J. Chim. Phys. 56 214 (1959); Electrochim. Acta 1 58 (1959)
M34 W. L. Marshall and E. U. Franck Report 1979, COV-79025-2 (Oak Ridge Natl. Lab., Tennessee)
M35 W. L. Marshall and E. V. Jones J. Phys. Chem. 70 4028 (1966)
M36 W. L. Marshall and R. Slusher J. Inorg. Nucl. Chem. 37 1191 (1975)
M37 A. E. Martell and G. Schwarzenbach Helv. Chim. Acta 39 653 (1956)
M38 F. S. Martin J. Chem. Soc. 1954 2564
M39 R. B. Martin and L. P. Henkle J. Phys. Chem. 68 3438 (1964)
M40 O. I. Martynova, Yu. F. Samoilov and T. I. Petrova Tr. Mosk. Energy. Inst. 238 66 (1975)
M41 O. I. Martynova, L. G. Vasina and S. A. Pozdnyckova Dokl. Akad. Nauk S.S.S.R. 202 1337 (1972)
M42 N. C. Marziano, P. G. Traverso, A. DeSantis and M. Sampoli J. Chem. Soc. Chem. Comm., 1978 873
M43 C. M. Mason and J. B. Culvern J. Am. Chem. Soc. 71 2387 (1949)
M44 J. G. Mason and A. D. Kowalak Inorg. Chem. 3 1248 (1964)
M45 M. R. Masson J. Inorg. Nucl. Chem. 38 545 (1976)
M46 S. Mateo and F. Brito An. Quim. 68 37 (1970)

M47 S. Mateo, A. Diaz anf F. Brito An. Quim. 67 1179 (1971)
M48 E. Matijević, J. P. Couch and M. Kerker J. Phys. Chem. 66 111 (1962)
M48a H. Matsuda and Y. Ayabe Z. Electrochem. 63 1164 (1959)
M48b H. Matsui and H. Ohtaki Bull. Chem. Soc., Japan, 47 2603 (1974)
M49 G. Mattock J. Am. Chem. Soc. 76 4835 (1954)
M50 M. Mavrodin-Tarabic Rev. Roum. Chem. 18 609 (1973)
M51 M. R. Mefod'eva, N. O. Krot, T. V. Afanes'ova and A. D. Gel'man Izv. Akad. Nauk. Ser. Khim. 10 2370 (1974)
M52 J. Meier and G. Schwarzenbach Helv. Chim. Acta 40 907 (1957)
M53 L. Meites J. Am. Chem. Soc. 75 6059 (1953)
M54 M. R. Melardi, G. Ferroni and J. Galea Bull. Soc. Chim. France 1976 1004
M55 H. Menard, J. P. Masson, J. Devynck and B. Tremillon J. Electroanal. Chem. Interfacial Electrochem. 63 163 (1975)
M56 H. Menzel Z. Physik. Chem. 100 276 (1922)
M57 H. Menzel Z. Physik. Chem. 105 402 (1923)
M58 E. E. Mercer and D. T. Farrar Canad. J. Chem. 46 2678 (1968)
M59 R. E. Mesmer Inorg. Chem. 10 857 (1971)
M60 R. E. Mesmer and C. F. Baes Inorg. Chem. 6 1951 (1967)
M61 R. E. Mesmer and C. F. Baes Inorg. Chem. 10 2290 (1971)
M62 R. E. Mesmer and C. F. Baes J. Solution Chem. 3 307 (1974)
M63 R. E. Mesmer, C. F. Baes and F. H. Sweeton J. Phys. Chem. 74 1937 (1970)
M64 R. E. Mesmer, C. F. Baes and F. H. Sweeton Inorg. Chem. 11 537 (1972)
M65 H. Metivier Report 1973. CEA—R—4477
M66 H. Metivier and R. Guillaumont Radiochem. Radioanal. Lett. 10 27 (1972)
M67 H. Metivier and R. Guillaumont Radiokhimiya 17 636 (1975)
M68 H. Metivier and R. Guillaumont Proc. Moscow Symp. Chem. Transuranium Elements 1972 179, Ed. V. Spitsyn and J. S. Katz, Pergamon Press, Oxford, 1976
M69 O. Meyerhof and K. Lohmann Biochem. Z. 196 22 (1928)
M70 L. Michaelis and T. Garmendia Biochem. Z. 67 431 (1914)
M71 L. Michaelis and M. Mizutani Z. Physik. Chem. 116 135 (1925)
M72 W. Miedrich, Thesis, Frankfurt-am-Main, 1954
M73 R. M. Milburn J. Am. Chem. Soc. 79 537 (1957)
M74 R. M. Milburn and W. C. Vosburgh J. Am. Chem. Soc. 77 1352 (1955)
M75 N. B. Milic Acta Chem. Scand. 25 2487 (1971)
M76 N. B. Milic, Z. D. Bugarcic and M. V. Vasic Glas. Hem. Drus. Beograd 45 349 (1980)
M77 F. J. Millero Geochim. Cosmochim. Acta 43 1651 (1979)
M78 J. K. Mishra and Y. K. Gupta Indian J. Chem. 6 757 (1968)
M79 H. C. Mishra and M. C. R. Symons J. Chem. Soc. 1962 1194
M80 A. D. Mitchell J. Chem. Soc. 117 957 (1920)
M81 A. G. Mitchell and W. F. K. Wynne-Jones Trans. Faraday Soc. 51 1690 (1955)
M82 L. I. Mit'kina, N. V. Mel'chakova and V. M. Peshkova Zhur. Neorg. Khim. 23 1258 (1978)

M83 R. P. Mitra, H. C. Malhotra and D. V. S. Jain Trans. Faraday Soc. 62 167 (1966)

M84 R. A. Mitra and B. R. Thakral Indian J. Chem. 8 347 (1970)

M85 T. Moeller J. Am. Chem. Soc. 63 1206 (1941)

M86 T. Moeller J. Am. Chem. Soc. 64 953 (1942)

M87 T. Moeller J. Phys. Chem. 50 242 (1946)

M88 T. Moeller and G. L. King J. Phys. Chem. 54 999 (1950)

M89 Y. V. Moiseev and M. I. Vinnik Doklady Akad. Nauk S.S.S.R. 150 845 (1963)

M90 J. M. Monger and O. Redlich J. Phys. Chem. 60 797 (1956)

M91 C. B. Monk J. Chem. Soc. 1949 423

M92 C. B. Monk and M. F. Amira J. Chem. Soc., Faraday Trans. I, 74 1170 (1978)

M93 T. S. Moore J. Chem. Soc. 91 1379 (1907)

M94 T. S. Moore and T. F. Winmill J. Chem. Soc. 101 1635 (1912)

M95 O. M. Morgan and O. Maass Canad. J. Res. 5 162 (1931)

M96 T. D. B. Morgan, G. Stedman and P. A. E. Whincup J. Chem. Soc. 1965 4813

M97 J. C. Morris J. Phys. Chem. 70 3798 (1966)

M98 C. Morton J. Chem. Soc. 1928 1401

M99 C. Morton Quart. J. Pharm. Pharmacol. 3 438 (1930)

M100 A. I. Moskvin Radiokhimiya 13 681 (1971)

M101 A. I. Moskvin and V. P. Zaitseva Radiokhimiya 4 73 (1962)

M102 A. Moutte, Thesis, University Paris, 1968; quoted by R. Guillaumont et al., C.R. Acad. Sci. (Paris), Ser. C 268 140 (1968)

M103 E. L. Muetterties, J. H. Balthis, Y. T. Chia, W. H. Knoth and H. C. Miller Inorg. Chem. 3 444 (1964)

M104 S. S. Muhammad and T. N. Rao J. Chem. Soc. 1957 1077

M105 S. S. Muhammad and T. N. Rao J. Indian Chem. Soc. 34 250 (1957)

M106 S. S. Muhammad, D. H. Rao and M. A. Haleem J. Indian Chem. Soc. 34 101 (1957)

M107 S. S. Muhammad and E. V. Sundaram J. Sci. Ind. Research (India) 20B 16 (1961)

M108 J. N. Mukerjee and B. Chatterjee Nature 155 85 (1945)

M109 L. N. Mulay and P. W. Selwood J. Am. Chem. Soc. 77 2693 (1955)

M110 H. D. Murray J. Chem. Soc. 127 882 (1925)

M111 Y. Musants and M. Porthault Radiochem. Radioanal. Lett. 15 299 (1973)

M112 J. Muus Z. Physik. Chem. 159A 268 (1932)

M113 L. T. Muus, I. Refn and R. W. Asmussen Trans. Danish Acad. Tech. Sci. No. 23 (1951)

M114 R. A. Myers Thesis, University of Nebraska, 1958

N

N1 B. I. Nabivanets Russ. J. Inorg. Chem. (Engl. Transl.) 7 212 (1962)

N2 B. I. Nabivanets Zhur. Neorg. Khim. 14 653 (1969)

N3 B. I. Nabivanets, E. F. Kapantsyan and E. N. Ogantsyan Zhur. Neorg. Khim. 19 729 (1974)

N4 B. I. Nabivanets and L. N. Kudritskaya Ukr. Khim. Zh. 30 891 (1964); CA 62 2290g (1965)

N5 B. I. Nabivanets and V. V. Lukachina Ukr. Khim. Zh. 30 1123 (1964); CA 62 7361h (1965)

N6 E. Nachbaur Monatsh. 91 749 (1960)

N7 S. Naidich and J. E. Ricci J. Am. Chem. Soc. 61 3268 (1939)

N8 V. S. K. Nair J. Inorg. Nucl. Chem. 26 1911 (1964)

N9 V. S. K. Nair and G. H. Nancollas J. Chem. Soc. 1958 4144

N10 R. Näsänen Suomen Kem. 19B 90 (1946)

N11 R. Näsänen Acta Chem. Scand. 1 204 (1947)

N12 R. Näsänen Acta Chem. Scand. 8 1587 (1954)

N13 R. Näsänen Suomen Kem. 33B 47 (1960)

N14 R. Näsänen and P. Meriläinen Suomen Kem. 33B 149 (1960)

N15 R. Näsänen and P. Meriläinen Suomen Kem. 33B 197 (1960)

N16 R. Näsänen, P. Meriläinen and K. Leppanen Acta Chem. Scand. 15 913 (1961)

N17 A. W. Naumann and C. J. Hallada Inorg. Chem. 3 70 (1964)

N18 V. A. Nazarenko, V. P. Antonovich and E. M. Nevskaya Zhur. Neorg. Khim. 13 1574 (1968)

N19 V. A. Nazarenko, V. P. Antonovich and E. M. Nevskaya Zhur. Neorg. Khim. 16 1804 (1971)

N20 V. A. Nazarenko, V. P. Antonovich and E. M. Nevskaya Zhur. Neorg. Khim. 16 2387 (1971)

N21 V. A. Nazarenko, V. P. Antonovich, A. P. Rubel and E. A. Biryuk Zhur. Neorg. Khim. 23 1787 (1978)

N22 V. A. Nazarenko and E. A. Biryuk, Zhur. Neorg. Khim. 19 632 (1974)

N23 V. A. Nazarenko and G. V. Flyantikova, Zhur. Neorg. Khim. 13 1855 (1968)

N24 V. A. Nazarenko and E. M. Nevskaya Zhur. Neorg. Khim. 14 3215 (1969)

N25 V. A. Nazarenko, G. G. Shitareva and E. N. Poluektova Zhur. Neorg. Khim. 18 1155 (1973)

N26 V. A. Nazarenko, G. G. Shitareva and E. N. Poluektova Zhur. Neorg. Khim. 22 980 (1977)

N27 B. V. Nekrasov and G. V. Zotov Zhur. Priklad. Khim. 14 264 (1941)

N28 G. Neumann Acta Chem. Scand. 18 278 (1964)

N29 G. Neumann Arkiv Kemi. 32 229 (1970)

N30 R. C. Neumann, W. Kauzmann and A. Zipp J. Phys. Chem. 77 2687 (1973)

N31 H. Neumann, I. Z. Steinberg and E. Katchalski J. Am. Chem. Soc. 87 3841 (1965)

N32 J. D. Neuss and W. Rieman J. Am. Chem. Soc. 56 2238 (1934)

N33 E. Newbery Trans. Electrochem. Soc. 69 611 (1936)

N34 L. Newman and D. N. Hume J. Am. Chem. Soc. 81 5901 (1959)

N35 L. Newman, W. J. Lafleur, F. J. Brousaides and A. M. Ross J. Am. Chem. Soc. 80 4491 (1958)

N36 R. F. Newton and M. G. Bolinger J. Am. Chem. Soc. 52 921 (1930)

N37 Nguyen Dinh Ngo and K. A. Burkov Zhur. Neorg. Khim. 19 1249 (1974)

N38 N. M. Nikolaeva Izv. Sib. Otd. Akad. Nauk S.S.S.R., Ser. Khim. Nauk. 1971 61

N39 N. M. Nikolaeva Izv. Sib. Otd. Akad. Nauk S.S.S.R. Ser. Khim. Nauk **1978** 91

N40 L. F. Nims J. Am. Chem. Soc. **55** 1946 (1933)

N41 L. F. Nims J. Am. Chem. Soc. **56** 1110 (1934)

N42 T. Nishide and R. Tsuchiya Bull. Chem. Soc. Japan **38** 1398 (1965)

N43 B. Noren Acta Chem. Scand. **27** 1369 (1973)

N44 A. V. Novoselova Zhur. Obshchei Khim. **1** 668 (1931)

N45 A. A. Noyes Carnegie Inst. Publ. **No. 63** (1907)

N46 A. A. Noyes, Y. Kato and R. B. Sosman J. Am. Chem. Soc. **32** 159 (1910)

N47 A. A. Noyes, Y. Kato and R. B. Sosman Z. Physik. Chem. **73** 1 (1910)

N48 P. Nylén Z. Anorg. Allgem. Chem. **230** 385 (1937)

N49 C. J. Nyman, D. K. Roe and R. A. Plane J. Am. Chem. Soc. **83** 323 (1961)

O

O1 S. J. O'Brien and E. G. Bobalek, J. Am. Chem. Soc. **62** 3227 (1940)

O2 P. R. O'Connor U.S. Atomic Energy Commission Report **CN-2033** (1944)

O3 H. G. Offner and D. A. Skoog Anal. Chem. **38** 1520 (1966)

O4 H. Ohtaki Acta Chem. Scand. **18** 521 (1964)

O5 H. Ohtaki Inorg. Chem., **6** 808 (1967)

O6 H. Ohtaki Inorg. Chem. **7** 1205 (1968)

O7 H. Ohtaki and G. Biederman Bull. Chem. Soc. Japan **44** 1822 (1971)

O8 H. Ohtaki and H. Kato Inorg. Chem. **6** 1935 (1967)

O9 H. Ohtaki and T. Kawai Bull. Chem. Soc. Japan **45** 1735 (1972)

O10 Y. Oka, K. Kawagaki and R. Kadoya J. Chem. Soc. Japan **64** 718 (1943)

O11 A. Ölander Z. Physik. Chem. **129** 1 (1927)

O12 A. Ölander Z. Physik. Chem. **144** 49 (1929)

O13 A. Olin Acta Chem. Scand. **11** 1445 (1957)

O14 A. Olin Acta Chem. Scand. **13** 1791 (1959)

O15 A. Olin Acta Chem. Scand. **14** 814 (1960)

O16 E. Olivieri-Mandala Gazz. Chim. Ital. **46** **1** 298 (1916)

O17 G. Olofsson J. Chem. Thermodyn. **7** 507 (1975)

O18 A. R. Olson and T. R. Simonson J. Chem. Phys. **17** 348, 1322 (1949)

O19 J. W. Olver and D. N. Hume Anal. Chim. Acta **20** 559 (1959)

O20 L. Onsager Physik. Z. **28** 277 (1927)

O21 S. M. Osinska-Tanevska, M. K. Bynyaeva, K. P. Mishchenko and I. E. Flis Russ. J. Appl. Chem. (Engl. transl.) **36** 1162 (1963)

O22 R. K. Osterheld J. Phys. Chem. **62** 1133 (1958)

O23 B. B. Owen J. Am. Chem. Soc. **56** 1695 (1934)

O24 B. B. Owen J. Am. Chem. Soc. **56** 2785 (1934)

O25 B. B. Owen J. Am. Chem. Soc. **57** 1526 (1935)

O26 B. B. Owen and E. J. King J. Am. Chem. Soc. **65** 1612 (1943)

P

P1 M. Paabo and R. G. Bates J. Phys. Chem. **73** 3014 (1969)

P2 M. Paabo, R. G. Bates and R. A. Robinson J. Phys. Chem. **70** 247 (1966)

P3 F. M. Page J. Chem. Soc. 1953 1719

P4 L. Pajdowski Roczniki Chem. 37 1351, 1363 (1963)

P5 L. Pajdowski and A. Olin Acta Chem. Scand. 16 983 (1962)

P6 V. V. Pal'chevskii Vestn. Leningrad. Univ., Fiz. Khim. 1972 144

P7 V. V. Pal'chevskii, T. I. L'vova and V. G. Krunchak Issled. Otd. Proized. Polufabrikat., Ochetki Prom. Stokon, 1972 166

P8 V. V. Pal'chevskii, T. I. L'vova and V. G. Krunchak, Tr. Vses. Nauch-Issled Inst. Tsellyul-Bum. Prom. 61 166 (1972)

P9 M. Palfalvi-Rozsahezyi, Z. G. Szabo and L. Barcza Acta. Chim. Acad. Sci. Hung. 104 303 (1980)

P10 A. V. Pamfilov and A. L. Agafonova Zhur. Fiz. Khim. 24 1147 (1950)

P11 K. Pan and T. M. Hseu Bull. Chem. Soc. Japan 28 162 (1955)

P12 A. J. Panson J. Phys. Chem. 67 2177 (1963)

P13 P. Paoletti, J. H. Stern and A. Vacca J. Phys. Chem. 69 3759 (1965)

P14 M. R. Paris and C. Gregoire Anal. Chem. Acta 42 431 (1968)

P15 M. R. Paris and C. Gregoire Anal. Chim. Acta 42 439 (1968)

P16 R. A. Paris and J. C. Merlin I.U.P.A.C. Inorg. Chem. Colloquium, Münster 1954 237

P17 P. R. Patel, E. C. Moreno and J. M. Patel J. Res. Nat. Bur. Stand., Sec. A 75 205 (1971)

P18 T. Paul Chem. Zeit. 23 535 (1899)

P19 M. A. Paul J. Am. Chem. Soc. 76 3236 (1954)

P20 M. A. Paul and F. A. Long Chem. Rev. 57 1 (1957)

P21 A. J. Paulson and D. R. Kosker J. Solution Chem. 9 269 (1980)

P22 L. A. Pavlyuk and P. A. Kryukov Izv. Sib. Otd. Akad. Nauk S.S.S.R., Ser. Khim. Nauk. 6 88 (1978)

P23 C. J. Peacock and G. Nickless Z. Naturforsch., A 24 245 (1969)

P24 D. Pearson, C. S. Copeland and S. W. Benson J. Am. Chem. Soc. 85 1047 (1963)

P25 R. G. Pearson and F. V. Williams J. Am. Chem. Soc. 76 258 (1954)

P26 R. L. Pecsok and A. N. Fletcher Inorg. Chem. 1 155 (1962)

P27 K. J. Pedersen Kgl. Danske Videnskab. Selskab. Math.-fys. Medd. 20 7 (1943)

P28 K. J. Pedersen Kgl. Danske Videnskab. Selskab. Math-fys. Medd. 22 10 (1945)

P29 L. Pentz and E. R. Thornton J. Am. Chem. Soc. 89 6931 (1967)

P30 V. D. Perkovets and P. A. Kryukov Izv. Sib. Otd. Akad. Nauk S.S.S.R., Ser. Khim. Nauk. 7 9 (1969)

P31 B. Perlmutter-Hayman and Y. Weisman J. Am. Chem. Soc. 91 668 (1969)

P32 C. Perrin J. Am. Chem. Soc. 86 256 (1964)

P33 D. D. Perrin J. Chem. Soc. 1959 1710

P34 D. D. Perrin J. Chem. Soc. 1960 3189

P35 D. D. Perrin J. Chem. Soc. 1962 2197

P36 D. D. Perrin J. Chem. Soc. 1962 4500

P37 D. D. Perrin J. Chem. Soc. 1964 3644

P38 D. Peschanski and J. M. Fruchart Compt. Rend. 257 1853 (1963)

P39 V. M. Peshkova and P. Ang Russ. J. Inorg. Chem. (Engl. transl.) **7** 1091 (1962)

P40 V. M. Peshkova, N. V. Mel'chakova and S. G. Zhemchuzhin Zhur. Neorg. Khim. **6** 1233 (1961)

P41 A. Peterson Acta Chem. Scand. **15** 101 (1961)

P42 A. D. Pethybridge and J. E. Prue Trans. Faraday Soc. **63** 2019 (1967)

P43 I.Z. Pevzner, N. I. Eremin, N. N. Kayazeva, Ya. E. Rozen and V. E. Mironov Zhur. Neorg. Khim. **18** 1129 (1973)

P44 J. N. Phillips Austral. J. Chem. **14** 183 (1961)

P45 R. Phillips, P. Eisenberg, P. George and R. J. Rutman J. Biol. Chem. **240** 4393 (1965)

P46 H. Pick Nernst Festschrift **1912** 374

P47 J. M. Pink Canad J. Chem., **48** 1169 (1970)

P48 A. L. Pitman, M. Pourbaix and N. de Zoubov J. Electrochem. Soc. **104** 594 (1957)

P49 V. A. Pleskov and A. M. Monoszon Zhur. Fiz. Khim. **6** 513 (1935)

P50 W. Plumb and G. M. Harris Inorg. Chem. **3** 542 (1964)

P51 J. J. Podestà Rev. Fac. Cienc. Quim. LaPlata **30** 61 (1957); CA **53** 14652 (1959)

P52 J. Podlaha and J. Podlahova Coll. Czech. Chem. Comm. **29** 3164 (1964)

P53 H. A. Pohl J. Chem. Eng. Data **6** 515 (1961)

P54 C. N. Polydoropoulos Chim. Chronika **24** 147 (1959)

P55 C. N. Polydoropoulos and M. Pipinis Z. Physik. Chem. (Frankfurt) **40** 322 (1964)

P56 N. K. Pongi, G. Double and J. Hurwic Bull. Soc. Chim. France **1980** 347

P57 G. Popa, C. Luca and E. Josif Z. Physik. Chem. **222** 49 (1963)

P58 C. Postmus and E. L. King J. Phys. Chem. **59** 1208 (1955)

P59 B. G. Pozharskii, T. N. Sterlingova and A. E. Petrova Russ. J. Inorg. Chem. (Engl. transl.) **8** 831 (1963)

P60 C. M. Preston and W. A. Adams J. Phys. Chem. **83** 814 (1979)

P61 E. B. R. Prideaux J. Chem. Soc. **99** 1224 (1911)

P62 E. B. R. Prideaux and A. T. Ward J. Chem. Soc. **125** 69 (1924)

P63 E. B. R. Prideaux and A. T. Ward J. Chem. Soc. **125** 425 (1924)

P64 M. Prytz Z. Anorg. Allgem. Chem. **174** 355 (1928)

P65 M. Prytz Z. Anorg. Allgem. Chem. **200** 133 (1931)

P66 W. Pugh J. Chem. Soc. **1929** 1994

P67 G. I. Pusenok and K. P. Mishchenko Zhur. Priklad. Khim. (Leningrad) **44** 934 (1971)

Q

Q1 D. Quane and J. E. Earley J. Am. Chem. Soc. **87** 3823 (1965)

Q2 M. Quintin J. Chim. Phys. **24** 712 (1927)

Q3 M. Quintin Compt. Rend. **204** 968 (1937)

Q4 M. Quintin Compt. Rend. **210** 625 (1940)

Q5 A. S. Quist J. Phys. Chem. **74** 3396 (1970)

Q6 A. S. Quist and W. L. Marshall J. Phys. Chem. **72** 3122 (1968)

Q7 A. S. Quist, W. L. Marshall and H. R. Jolley J. Phys. Chem. 69 2726 (1965)

R

R1 J. Rabani and M. S. Matheson J. Am. Chem. Soc. 86 3175 (1964)
R2 J. Rabani, W. A. Mulac and M. S. Matheson J. Phys. Chem. 69 53 (1965)
R3 S. W. Rabideau J. Am. Chem. Soc. 79 3675 (1957)
R4 S. W. Rabideau, unpubl., quoted by T. K. Keenan J. Phys. Chem. 61 1117 (1957)
R5 S. W. Rabideau and R. J. Kline J. Phys. Chem. 62 617 (1958)
R6 S. W. Rabideau and R. J. Kline J. Phys. Chem. 64 680 (1960)
R7 S. W. Rabideau and J. F. Lemons J. Am. Chem. Soc. 73 2895 (1951)
R8 A. J. Rabinowitsch and E. Laskin Z. Physik. Chem. A134 387 (1928)
R9 R. Radegliu Z. Physik. Chem. (Leipzig) 231 239 (1966)
R10 A. T. Ram, D. Ramaswamy and Y. Nayudamma Recent Advan. Mineral Tannages, Papers Symp., Madras 1964 207; CA 64 18510 (1966)
R11 R. W. Ramette and R. F. Stewart J. Phys. Chem. 65 243 (1961)
R12 S. R. Rao and L. G. Hepler Hydrometallurgy 2 293 (1977)
R13 T. Rathlev and T. Rosenberg Arch. Biochem. Biophys. 65 319 (1956)
R14 P. C. Ray, M. L. Dey and J. C. Ghosh J. Chem. Soc. 111 413 (1917)
R15 O. Redlich Z. Physik. Chem. A182 42 (1938)
R16 O. Redlich and J. Bigeleisen J. Am. Chem. Soc. 65 1883 (1943)
R17 O. Redlich, R. W. Duerst and A. Merback J. Chem. Phys. 49 2986 (1968)
R18 A. J. Reed J. Solution Chem. 4 53 (1975)
R19 C. A. Reynolds and W. J. Argersinger J. Phys. Chem. 56 417 (1952)
R20 W. L. Reynolds and S. Fukushima Inorg. Chem. 2 176 (1963)
R21 D. H. Richards and K. W. Sykes J. Chem. Soc. 1960 3626
R22 P. H. Rieger Austral. J. Chem. 26 1173 (1973)
R23 A. I. Rivkind Doklady Akad. Nauk S.S.S.R. 142 137 (1962)
R24 D. Roach and E. S. Amis Z. Physik. Chem. (Frankfurt) 35 274 (1962)
R25 E. J. Roberts J. Am. Chem. Soc. 52 3877 (1930)
R26 E. B. Robertson and H. B. Dunford J. Am. Chem. Soc. 86 5080 (1964)
R27 T. J. Robertson and E. Hartwig Canad. J. Chem. 29 818 (1951)
R28 R. A. Robinson Trans. Faraday Soc. 32 743 (1936)
R29 R. A. Robinson and R. G. Bates, Anal. Chem. 43 969 (1971)
R30 R. A. Robinson and V. E. Bower J. Phys. Chem. 65 1279 (1961)
R31 T. E. Rodgers and G. M. Waind Trans. Faraday Soc. 57 1360 (1961)
R32 E. F. C. H. Rohwer, J. A. Brink and J. J. Cruywagen J. S. Afr. Chem. Soc. 28 1 (1975)
R33 E. F. C. H. Rohwer and J. J. Cruywagen J. S. African Chem. Inst. 16 26 (1963)
R34 E. F. C. H. Rohwer and J. J. Cruywagen J. S. African Chem. Inst. 17 145 (1964)
R35 A. Rosenheim and G. Jander Kolloid Z. 22 23 (1918)
R36 A. Rosenheim and L. Krause Z. Anorg. Allgem. Chem. 118 177 (1921)

R37 V. Ya. Rosolovskii and I. U. Kolesnikov Zhur. Neorg. Khim. 13 1801 (1968)

R38 W. H. Ross Proc. Trans. Nova Scotian Inst. Sci. 11 95 (1905)

R39 S. D. Ross and A. J. Catotti J. Am. Chem. Soc. 71 3563 (1949)

R40 D. R. Rosseinsky and M. J. Nicol J. Chem. Soc., A 1968 1022

R41 F. J. C. Rossotti and H. S. Rossotti Acta Chem. Scand. 9 1177 (1955)

R42 F. J. C. Rossotti and H. S. Rossotti Acta Chem. Scand. 10 779 (1956)

R43 F. J. C. Rossotti and H. S. Rossotti Acta Chem. Scand. 10 957 (1956)

R44 W. A. Roth Annalen 542 35 (1939)

R45 W. A. Roth and E. Börger Z. Elektrochem. 43 354 (1937)

R46 W. A. Roth and O. Schwarz Ber. 59 338 (1926)

R47 V. Rothmund and K. Drucker Z. Physik. Chem. 46 827 (1903)

R48 K. Rothschein, J. Socha, P. Vetesnik and M. Vecera Coll. Czech. Chem. Comm. 35 3128 (1970)

R49 F. J. W. Roughton J. Am. Chem. Soc. 63 2930 (1941)

R50 J. K. Ruff Inorg. Chem. 4 1446 (1965)

R51 C. K. Rule and V. K. LaMer J. Am. Chem. Soc. 60 1974 (1938)

R52 C. I. Rulfs, R. F. Hirsch and R. A. Pacer Nature 199 96 (1963)

R53 P. Rumpf Compt. Rend. 197 686 (1933)

R54 P. Rumpf and V. Chavane Compt. Rend 224 919 (1947)

R55 R. M. Rush and J. S. Johnson J. Phys. Chem. 67 821 (1963)

R56 R. M. Rush, J. S. Johnson and K. A. Kraus Inorg. Chem. 1 378 (1962)

R57 R. S. Ryabova, I. M. Medvetskaya and M. I. Vinnik Zhur. Fiz. Khim. 40 339 (1966)

R58 J. Rydberg Arkiv Kemi 8 113 (1955)

R59 S. Rydholm Svensk. Papperstidn. 58 273 (1955)

R60 I. G. Ryss and V. B. Tul'chinskii Zhur. Neorg. Khim. 6 1856 (1961)

R61 B. N. Ryzhenko Doklady Akad. Nauk. S.S.S.R. 149 639 (1963)

R62 B. N. Ryzhenko Geokhimiya 1963 137

R63 B. N. Ryzhenko Geokhimiya 1965 273

R64 B. N. Ryzhenko Geokhimiya 1967 161

S

S1 R. N. J. Saal Rec. Trav. Chim. 47 264 (1928)

S2 R. Sabbah and G. Carpeni J. Chim. Phys. 63 1549 (1966)

S3 M. A. Salam and M. A. Raza Chem. Ind. (London) 1971 601

S4 K. Sallavo and P. Lumme Suomen Kemistilehti B 40 155 (1967)

S5 K. Y. Salnis, K. P. Mishchenko and I. E. Flis Zhur. Neorg. Khim. 2 1985 (1957)

S6 P. Salomaa, R. Hakala, S. Vesala and T. Aalto Acta Chem. Scand. 23 2116 (1969)

S7 P. Salomaa, L. L. Schaleger and F. A. Long J. Phys. Chem. 68 410 (1964)

S8 P. Salomaa, L. L. Schaleger and F. A. Long J. Am. Chem. Soc. 86 1 (1964)

S9 P. Salomaa and A. Vasala Acta Chem. Scand. 20 1414 (1966)

S10 P. Salomaa, A. Vesala and S. Vesala Acta Chem. Scand. 23 2107 (1969)

S11 J. Sand Z. Physik. Chem. 48 614 (1904)

S12 Y. I. Sannikov, V. I. Zolotavin and I. Y. Bezrukov Russ. J. Inorg. Chem (Engl. transl) **8** 474 (1963)

S13 R. S. Sapieszko, R. C. Patel and E. Matijevic J. Phys. Chem. **81** 1061 (1977)

S14 P. L. Sarma and M. S. Davis J. Inorg. Nucl. Chem. **29** 2607 (1967)

S15 Y. Sasaki Acta Chem. Scand. **15** 175 (1961)

S16 Y. Sasaki Acta Chem. Scand. **16** 719 (1962)

S17 Y. Sasaki, I. Lindqvist and L. G. Sillén J. Inorg. Nucl. Chem. **9** 93 (1959)

S18 Y. Sasaki and L. G. Sillén Acta Chem. Scand. **18** 1014 (1964)

S19 Y. Sasaki and L. G. Sillén Arkiv Kemi **29** 253 (1968)

S20 D. P. N. Satchell J. Chem. Soc. **1958** 3904

S21 R. D. Sauerbrunn and E. B. Sandell J. Am. Chem. Soc. **75** 4170 (1953)

S22 A. P. Savostin Zhur. Neorg. Khim. **11** 2817 (1966)

S23 A. J. Scarpone Diss. Abs. Int. B **30** 1027 (1969)

S24 R. Schaal and P. Favier Bull. Soc. Chim. France **1959** 2011

S25 U. Schedin Acta Chem. Scand. **25** 747 (1971)

S26 U. Schedin Acta Chem. Scand., Ser. A **29** 333 (1975)

S27 U. Schedin and M. Frydman Acta Chem. Scand. **22** 115 (1968)

S28 P. G. Scheurer, R. M. Brownell and J. E. LuValle J. Phys. Chem. **62** 809 (1958)

S29 K. Schiller and E. Thilo Z. Anorg. Allgem. Chem. **310** 261 (1961)

S30 H. Schmid, R. Marchgraber and F. Dunkl Z. Elektrochem. **43** 337 (1937)

S31 L. Schoepp Diss. Techn. Univ. Berlin, **1960**

S32 R. K. Schofield and A. W. Taylor J. Chem. Soc. **1954** 4445

S33 N. H. Schoon and L. Wannholt Svensk. Papperstidn. **72** 431 (1969)

S34 G. Schorsch Thesis, Univ. Strasburg, **1961**

S35 G. Schorsch Bull. Soc. Chim. France **1964** 1449

S36 G. Schorsch Bull. Soc. Chim. France **1964** 1456

S37 G. Schorsch Bull. Soc. Chim. France **1965** 988

S38 G. Schorsch and J. Byé Compt. Rend. **257** 2833 (1963)

S39 G. Schorsch and N. Ingri Acta Chem. Scand. **21** 2727 (1967)

S40 B. Schrager Collection Czech. Chem. Commun. **1** 275 (1929)

S41 E. Schreiner Z. Physik. Chem. **111** 419 (1924)

S42 M. Schumann Ber. **33** 527 (1900)

S43 G. M. Schwab Chem. Ber. **90** 221 (1957)

S44 G. M. Schwab and K. Polydoropoulos Z. Anorg. Allgem. Chem. **274** 234 (1953)

S45 R. Schwarz and E. Huf Z. Anorg. Allgem. Chem. **203** 188 (1931)

S46 R. Schwarz and W. D. Müller Z. Anorg. Allgem. Chem. **296** 273 (1958)

S47 G. Schwarzenbach Helv. Chim. Acta **19** 178 (1936)

S48 G. Schwarzenbach Z. Physik. Chem. **176** 133 (1936)

S49 G. Schwarzenbach Pure Appl. Chem. **5** 377 (1962)

S50 G. Schwarzenbach, A. Epprecht and H. Erlenmeyer Helv. Chim. Acta **19** 1292 (1936)

S51 G. Schwarzenbach and G. Geier Helv. Chim. Acta **46** 906 (1963)

S52 G. Schwarzenbach, G. Geier and J. Littler Helv. Chim. Acta 45 2601 (1962)

S53 G. Schwarzenbach and R. Sulzberger Helv. Chim. Acta 27 348 (1944)

S54 G. Schwarzenbach and J. Zurc Monatsh. 81 202 (1950)

S55 F. Seel and R. Schwaebel Z. Anorg. Allgem. Chem. 274 169 (1953)

S56 F. Seel and R. Winkler Z. Physik. Chem. (Frankfurt) 25 217 (1960)

S57 T. Sekine Acta Chem. Scand. 19 1526 (1965)

S58 T. Sekine, H. Iwaki, M. Sakairi, F. Shimada and M. Inarida Bull. Chem. Soc. Japan 41 1 (1968)

S59 J. Sendroy and A. B. Hastings J. Biol. Chem. 71 785 (1927)

S60 G. M. Sergeev and V. D. Almazova Tr. Khim. Khim. Tekhnol. 1970 31

S61 E. P. Sevostyanova and G. V. Khalturin Radiokhimiya 18 870 (1976)

S62 T. M. Seward Geochim. Cosmochim. Acta 38 1651 (1974)

S63 J. Shankar and B. C. De Souza Austral. J. Chem. 16 1119 (1963)

S64 L. Sharma and B. Prasad J. Indian Chem. Soc. 46 241 (1969)

S65 S. A. Shcherbakova, N. A. Krasnyanskaya, N. W. Mel'chakova and V. M. Peshkova Zhur. Neorg. Khim 23 770 (1978)

S66 T. Shedlovsky and D. A. MacInnes J. Am. Chem. Soc. 57 1705 (1935)

S67 M. S. Sherrill J. Am. Chem. Soc. 29 1641 (1907)

S68 M. S. Sherrill, C. B. King and R. C. Spooner J. Am. Chem. Soc. 65 170 (1943)

S69 M. S. Sherrill and A. A. Noyes J. Am. Chem. Soc. 48 1861 (1926)

S70 V. B. Shevcenko, I. V. Shilin and Yu. K. Zhdavov Zh. Neorg. Khim. 5 2832 (1960)

S71 E. A. Shilov Zhur. Fiz. Khim. 24 702 (1950)

S72 E. A. Shilov and J. N. Gladtchikova J. Am. Chem. Soc. 60 490 (1938)

S73 E. A. Shilov and N. P. Kanyaev Zhur. Fiz. Khim. 5 654 (1934)

S74 D. W. Shoesmith and W. Lee Canad. J. Chem. 54 3553 (1976)

S75 T. H. Siddell and W. C. Vosburgh J. Am. Chem. Soc. 73 4270 (1951)

S76 J. Sierra, M. Ojeda and P. H. Wyatt J. Chem. Soc., B 1970 1570

S77 M. D. Silverman and H. A. Levy Amer. Chem. Soc. 118th meeting, Chicago, Sept. 1950, 5Q.

S78 L. Silvester and P. A. Rock J. Chem. Eng. Data 19 98 (1974)

S79 K. Sima J. Biochem. (Japan) 29 121 (1939)

S80 C. R. Singleterry, Thesis, Univ. Chicago, 1940

S81 J. E. Singley Diss. Abs. B 27 3803 (1967)

S82 G. S. Sinyakova Latv. PSR Zinat. Akad. Vestn. Khim. Ser. 1975 330

S83 A. M. Sirota, R. A. Scheinin and I. I. Gol'dshtein Teploenergetika (Moscow) 1980 68

S83a L. H. Skibsted and P. C. Ford Acta Chem. Scand., Sec. A. A34 109 (1980)

S84 A. Skrabal Ber. 75B 1570 (1942)

S85 A. Skrabal and A. Berger Monatsh. 70 163 (1937)

S86 A. Skrabal and R. Skrabal Monatsh. 71 251 (1938)

S87 A. Skrabal and A. Zahorka Z. Elektrochem. 33 42 (1927)

S88 R. Smied J. Inorg. Nucl. Chem. 37 318 (1975)

S89 G. B. L. Smith, F. P. Gross, G. H. Brandes and A. W. Browne J. Am. Chem. Soc. 56 1116 (1934)

S90 S. J. Smrz Rec. Trav. Chim. 44 580 (1925)
S91 N. Soffer, Y. Marcus and J. Shamir J. Chem. Soc. Faraday Trans. I 76 2347 (1980)
S92 A. S. Solovkin Zhur. Neorg. Khim. 2 611 (1957)
S93 A. S. Solovkin and A. I. Ivantsov Zh. Neorgan. Khim. 11 1897 (1966)
S94 A. S. Solovkin, Z. N. Tsvetkova and A. I. Ivantsov Zhur. Neorg. Khim. 12 626 (1967)
S95 F. G. Soper J. Chem. Soc. 125 2227 (1924)
S96 Z. Soubarew-Chatelain Compt. Rend. 208 584 (1939)
S97 P. Souchay and G. Carpéni Bull. Soc. Chim. France 13 160 (1946)
S98 P. Souchay and A. Hessaby Bull. Soc. Chim. France 1953 614
S99 P. Souchay and R. Schaal Bull. Soc. Chim. France 17 819 (1950)
S100 P. Souchay and R. Schaal Bull. Soc. Chim. France 17 824 (1950)
S101 G. Sourisseau Ann. Chim. (Paris) [12] 8 349 (1953)
S102 M. Spiro Trans. Faraday Soc. 55 1746 (1959)
S103 E. Spitalsky Z. Anorg. Allgem. Chem. 54 265 (1907)
S104 V. B. Spivchovskii and L. P. Moisa Zhur. Neorg. Khim. 9 2287 (1964)
S105 K. Srinivasan and G. A. Rechnitz Anal. Chem. 40 1818 (1968)
S106 A. I. Stabrovskii Zhur. Fiz. Khim. 26 949 (1952)
S107 M. von Stackelberg and H. von Freyhold Z. Elektrochem. 46 120 (1940)
S108 W. C. Stadie and E. R. Hawes J. Biol. Chem. 77 241 (1928)
S109 J. Starý Collection Czech. Chem. Commun. 25 890 (1960)
S110 J. Ste-Marie, A. E. Torma and A. O. Gübeli Canad. J. Chem. 42 662 (1964)
S111 A. V. Stepanov and V. P. Shvedov Russ. J. Inorg. Chem. (Engl. transl.) 10 541 (1965)
S112 H. P. Stephens and J. W. Cobble Inorg. Chem. 10 619 (1971)
S112a R. Stewart and J. P. O'Donnell Canad. J. Chem. 42 1681 (1964)
S113 D. I. Stock and C. W. Davies Trans. Faraday Soc. 44 856 (1948)
S114 D. Stojkovic, M. M. Jacksic and B. Z. Nikolic Glas. Hem. Drus. Beograd 34 171 (1969)
S115 N. G. Stretenskaya Geochimiya 1977 430
S116 P. E. Sturrock, J. D. Ray, J. McDowell and H. R. Hunt Inorg. Chem. 2 649 (1963)
S117 T. N. Sudakova, A. A. Agoev, D. A. Denisov, V. V. Krasnoshchekov and Yu. G. Frolov Deposited Doc., 1977, VINITI 3433-77
S118 T. N. Sudakova and V. V. Krasnoshchekov Zhur. Neorg. Khim. 23 1506 (1978)
S119 T. N. Sudakova, V. V. Krasnoshchekov and Yu. G. Frolov Zhur. Neorg. Khim. 23 2092 (1978)
S120 J. C. Sullivan and J. C. Hindman J. Phys. Chem. 63 1332 (1959)
S121 N. Sundén Svensk. Kem. Tidskr. 66 20 (1954)
S122 N. Sundén Svensk. Kem. Tidskr. 66 50 (1954)
S123 L. H. Sutcliffe and J. R. Weber Trans. Faraday Soc. 52 1225 (1956)
S124 L. H. Sutcliffe and J. R. Weber J. Inorg. Nucl. Chem. 12 281 (1960)
S125 J. Sutton Nature 169 71 (1952)
S126 T. Suzuki and H. Hagisawa Bull. Inst. Phys. Chem. Research (Tokyo) 21 601 (1942)

S127 T. W. Swaddle and P. C. Kong Canad. J. Chem. 48 3223 (1970)

S128 K. Swaminathan and G. M. Harris J. Am. Chem. Soc. 88 4411 (1966)

S129 F. H. Sweeton and C. F. Baes J. Chem. Thermodynam. 2 479 (1970)

S130 F. H. Sweeton, R. E. Mesmer and C. F. Baes J. Solution Chem. 3 191 (1974)

S131 R. N. Sylva and P. L. Brown J. Chem. Soc., Dalton Trans., 1980 1577

S132 R. N. Sylva and M. R. Davidson J. Chem. Soc., Dalton Trans., 1979 232

S133 R. N. Sylva and M. R. Davidson J. Chem. Soc., Dalton Trans., 1979 465

S134 I. Szilard Acta Chem. Scand. 17 2674 (1963)

T

T1 K. Tachiki J. Chem. Soc. Japan 65 346 (1944)

T2 T. Tadeusz and S. Kiciak Rocz. Chem. 46 1209 (1972)

T3 K. Takahasi and N. Yui Bull. Inst. Phys. Chem. Research (Tokyo) 20 521 (1941)

T4 H. Takaya, W. Todo and T. Idosoya Bull. Chem. Soc. Japan 42 2748 (1969)

T5 H. V. Tartar and H. H. Garretson J. Am. Chem. Soc. 63 808 (1941)

T5a I. V. Tat'yanina, A. P. Borisova, E. A. Torchenkova and V. I. Spitsyn Dokl. Akad. Nauk S.S.S.R. 256 612 (1981)

T6 K. Täufel, C. Wagner and H. Dünwald Z. Elektrochem. 34 115 (1928)

T7 E. G. Taylor, R. P. Desch and A. J. Catotti J. Am. Chem. Soc. 73 74 (1951)

T8 B. J. Thamer J. Am. Chem. Soc. 79 4298 (1957)

T9 A. Thiel and H. Gessner Z. Anorg. Allgem. Chem. 86 1 (1914)

T10 L. C. Thompson, Thesis, Wayne Univ. Detroit, 1955

T11 R. S. Tobias Acta Chem. Scand. 12 198 (1958)

T12 R. S. Tobias J. Am. Chem. Soc. 82 1070 (1960)

T13 R. S. Tobias and A. B. Garrett J. Am. Chem. Soc. 80 3532 (1958)

T14 L. R. Tokareva and M. V. Mokhossoev Zhur. Neorg. Khim. 16 2417 (1971)

T15 V. N. Tolmachev and L. N. Serpukhova Zhur. Fiz. Khim. 30 134 (1956)

T16 J. Y. Tong Inorg. Chem. 3 1804 (1964)

T17 J. Y. Tong and R. L. Johnson Inorg. Chem. 5 1902 (1966)

T18 J. Y. Tong and E. L. King J. Am. Chem. Soc. 75 6180 (1953)

T19 L. K. J. Tong and A. R. Olson J. Am. Chem. Soc. 63 3406 (1941)

T20 H. Töpelmann J. Prakt. Chem. 121 320 (1929)

T21 I. Tossidis Inorg. Nucl. Chem. Lett. 12 609 (1976)

T22 A. R. Tourky and A. A. Mousa J. Chem. Soc. 1949 1305

T23 W. D. Treadwell and G. Schwarzenbach Helv. Chim. Acta 11 405 (1928)

T24 W. D. Treadwell and W. Wieland Helv. Chim. Acta 13 842 (1930)

T25 S. Tribalat and J. M. Caldero Bull. Soc. Chim. France 1966 774

T26 S. Tribalat and L. Schriver Bull. Soc. Chem. France 1975 2012

T27 R. Tsuchiya and A. Umavshara Bull. Chem. Soc. Japan 36 554 (1963)

T28 H. Tsukuda, T. Kawai, M. Maeda and H. Ohtaki Bull. Chem. Soc. Japan 48 691 (1975)

T29 T. A. Tumanova, K. P. Mishchenko and I. E. Flis Zhur. Neorg. Khim. 2 1990 (1957)

T30 J. Tummavuori and P. Lumme Acta Chem. Scand. 22 2003 (1968)
T31 D. J. Turner J. Chem. Soc., Faraday Trans. I 70 1346 (1974)
T32 R. C. Turner Canad. J. Chem. 53 2811 (1975)
T33 R. C. Turner and K. E. Miles Canad. J. Chem. 35 1002 (1937)

U

U1 N. Uri Chem. Rev. 50 375 (1952)
U2 L. N. Usherenko and N. A. Skorik Zhur. Neorg. Khim. 17 2918 (1972)

V

V1 A. Vaillant, J. Devynck and B. Tremillon Anal. Lett. 6 1095 (1973)
V2 N. E. Vanderborgh Talanta 15 1009 (1968)
V3 C. E. Vanderzee J. Am. Chem. Soc. 74 4806 (1952)
V4 C. E. Vanderzee and A. S. Quist J. Phys. Chem. 65 118 (1961)
V5 J. A. Van Lier, P. L. de Bruyn and J. T. G. Overbeek J. Phys. Chem. 64 1675 (1960)
V6 A. Vasala and V. Koskinen Finn. Chem. Lett. 5 145 (1975)
V7 E. G. Vassian and W. A. Eberhardt J. Phys. Chem. 62 84 (1958)
V8 V. M. Vdovenko, L. N. Lazarov and Yu. S. Khvorostin Radiokhimiya 9 460, 464 (1967)
V9 V. M. Vdovenko, L. N. Lazarov and Yu. S. Khvorostin Zhur. Neorg. Khim. 12 1152 (1967)
V10 V. M. Vdovenko, G. A. Romanov and V. A. Shcherbakov Radiokhimiya 5 137 (1963); CA 59 14639 (1963)
V11 A. Vertes, M. Ranogajec-Komor and P. Gelensev Acta Chim. (Budapest) 77 55 (1973)
V12 K. A. Vesterberg Z. Anorg. Allgem. Chem., 94 371 (1916)
V13 P. Vetesnik, J. Bielawsky and M. Vecera Coll. Czech. Chem. Comm. 33 1687 (1968)
V14 P. Vetesnik, K. Rothschein, J. Socha and M. Vecera Coll. Czech. Chem. Comm. 34 1087 (1969)
V15 J. Vilim Coll. Czech. Chem. Comm. 26 1268 (1961)
V16 M. I. Vinnik Usp. Khim. 35 1922 (1966)
V17 M. I. Vinnik, R. N. Kruglov and N. M. Chirkov Zhur. Fiz. Khim. 30 827 (1956)
V18 A. A. Vlcek Collection Czech. Chem. Commun. 20 400 (1955)
V19 Yu. A. Volokhov, L. N. Pavlov, N. I. Eremin and V. E. Mironov Zhur. Prikl. Khim. (Leningrad) 44 246 (1971)
V20 V. I. Volk and A. D. Bellkov Radiokhimiya 19 811 (1977)
V21 J. Vuceta and J. J. Morgan Limnol. Oceanograph. 22 742 (1977)

W

W1 J. Walker and W. Cormack J. Chem. Soc. 77 5 (1900)
W2 R. M. Wallace J. Phys. Chem. 70 3922 (1966)
W3 G. K. Ward and E. J. Millero J. Solution Chem. 3 417 (1974)

W4 G. C. Ware, J. B. Spulnik and E. C. Gilbert J. Am. Chem. Soc. 58 1605 (1936)

W5 L. J. Warren Anal. Chim. Acta 53 199 (1971)

W6 J. A. Wasastjerna Soc. Sci. Fennica Com. Phys. Mat 1 (1923)

W7 E. W. Washburn J. Am. Chem. Soc. 30 31 (1908)

W8 E. W. Washburn and E. K. Strachan J. Am. Chem. Soc. 35 681 (1913)

W9 S. Wasif J. Chem. Soc. A 1967 142

W10 J. I. Watters and A. Aaron J. Am. Chem. Soc. 75 611 (1953)

W11 J. I. Watters, E. D. Loughran and S. M. Lambert J. Am. Chem. Soc. 78 4855 (1956)

W12 J. I. Watters, P. E. Sturrock and R. E. Simonaitis Inorg. Chem. 2 765 (1963)

W13 O. Weider Ber. 68B 1423 (1935)

W14 J. L. Weeks and J. Rabani J. Phys. Chem. 70 2100 (1966)

W15 I. Weil and J. C. Morris J. Am. Chem. Soc. 71 3123 (1949)

W16 J. Weiss Trans. Faraday Soc. 31 966 (1935)

W17 R. C. Wells J. Washington Acad. Sci. 32 321 (1942)

W18 C. F. Wells and G. Davies Nature 205 692 (1965)

W19 C. F. Wells and M. A. Salam Nature 205 690 (1965)

W20 H. Wenger Dissertation, Eidg. Techn. Hochschule, Zurich, 1964

W21 C. A. West J. Chem. Soc. 77 705 (1900)

W22 R. E. Weston and J. Bigeleisen J. Am. Chem. Soc. 76 3074 (1954)

W23 E. J. Wheelwright, F. H. Spedding and G. Schwarzenbach J. Am. Chem. Soc. 75 4196 (1953)

W24 M. Whitfield J. Chem. Eng. Data 17 124 (1972)

W25 M. Widmer and G. Schwarzenbach Helv. Chim. Acta 47 266 (1964)

W26 J. S. Willcox and E. B. R. Prideaux J. Chem. Soc. 127 1543 (1925)

W27 A. S. Wilson and H. Taube J. Am. Chem. Soc. 74 3509 (1952)

W28 K. Winkelblech Z. Physik. Chem. 36 546 (1901)

W29 D. C. Winkley Diss. Abs. 26 7027 (1966)

W30 T. H. Wirth and N. Davidson J. Am. Chem. Soc. 86 4325 (1964)

W31 F. K. Wissbrun, D. M. French and A. Patterson J. Phys. Chem. 58 693 (1954)

W32 J. A. Wolhoff and J. T. G. Overbeek Red. Trav. Chim. 78 759 (1959)

W33 J. K. Wood J. Chem. Soc. 93 411 (1908)

W34 J. K. Wood J. Chem. Soc. 97 878 (1910)

W35 R. H. Wood J. Am. Chem. Soc. 80 1559 (1958)

W36 E. M. Woolley, J. O. Hill, W. K. Hannan and L. G. Hepler J. Solution Chem. 7 385 (1978)

W37 C. B. Wooster J. Am. Chem. Soc. 60 1609 (1938)

W38 M. Wrewski, N. Sawaritzki and L. Scharloff Z. Phys. Chem. 112 97 (1924)

W39 J. M. Wright, W. T. Lindsay and T. R. Druga U.S. Atomic Energy Comm. WAPD-TM-204 (1961)

W40 R. H. Wright and O. Maass Canad. J. Res. 6 588 (1932)

W41 P. A. H. Wyatt Discussions Faraday Soc. 24 162 (1957)

W42 P. A. H. Wyatt Tras. Faraday Soc. 56 490 (1960)

W43 W. F. K. Wynne-Jones J. Chem. Soc. 1930 1064

W44 W. F. K. Wynne-Jones Proc. Roy. Soc. A140 440 (1933)

W45 W. F. K. Wynne-Jones Trans. Faraday Soc. 32 1397 (1936)

Y

Y1 G. Yagil J. Phys. Chem. 71 1034, 1045 (1967)

Y2 G. Yagil and M. Anbar J. Am. Chem. Soc. 85 2376 (1963)

Y3 G. Yagil and M. Anbar J. Inorg. Nucl. Chem. 26 453 (1964)

Y4 Yu. B. Yakovlev, F. Ya. Kul'ba and D. A. Zenchenko Zhur. Neorg. Khim. 21 61 (1976)

Y5 Kh. M. Yakubov, E. Ya. Offengenden and V. V. Pal'chevski Kompleksobravokistelmo-vosstanovlist 1972 52

Y6 Kh. M. Yakubov and E. Ya. Offengenden Kompleksobravokistelmo-vosstanovlist 1972 60

Y7 S. Yamada, S. Funahashi and M. Tanaka, J. Inorg. Nucl. Chem., 37 835 (1975)

Y8 T. Yamane and N. Davidson J. Am. Chem. Soc. 82 2123 (1960)

Y9 K. Yates Canad. J. Chem. 42 1239 (1964)

Y10 K. Yates and J. C. Riordan Canad. J. Chem. 43 2328 (1965)

Y11 K. Yates and J. B. Stevens Canad. J. Chem. 43 529 (1965)

Y12 K. Yates, J. B. Stevens and A. R. Katritzky Canad. J. Chem. 42 1957 (1964)

Y13 K. Yates and H. Wai J. Am. Chem. Soc. 86 5408 (1964)

Y14 K. B. Yatsimirskii and I. I. Alekseeva Izvest. Vysshikh Ucheb. Zavedenii, Khim. i Khim. Tekhnol. No. 1 53 (1958)

Y15 K. B. Yatsimirskii and V. E. Kalinina Russ. J. Inorg. Chem. (Engl. transl.) 9 611 (1964)

Y16 K. R. Yatsimirskii and K. E. Prik Zhur. Neorg. Khim. 9 178 (1964)

Y17 K. R. Yatsimirskii and K. E. Prik Zhur. Neorg. Khim. 9 1838 (1964)

Y18 F. H. Yorston Pulp and Paper Mag. Canada 31 374 (1931)

Y19 D. M. Yost and R. J. White J. Am. Chem. Soc. 50 81 (1928)

Y20 T. F. Young, L. F. Maranville and H. M. Smith, unpubl., quoted by T. F. Young and D. E. Irish in Ann. Rev. Phys. Chem. 13 435 (1962)

Y21 N. Yui Bull. Inst. Phys. Chem. Research (Tokyo) 19 1229 (1940)

Y22 N. Yui Bull. Inst. Phys. Chem. Research (Tokyo) 20 256 (1941)

Y23 N. Yui Bull. Inst. Phys. Chem. Research (Tokyo) 20 390 (1941)

Y24 N. Yui Sci. Reports Tohoku Univ. 1st series 35 53 (1951)

Y25 N. Yui and H. Hagisawa Bull. Inst. Phys. Chem. Research (Tokyo) 21 597 (1942)

Z

Z1 S. S. Zavodnov and P. A. Kryukov Izvest. Akad. Nauk S.S.S.R. Otdel. Khim. Nauk 1960 1704

Z2 J. V. Zawidzky Ber. 36 1427 (1903)

Z3 E. L. Zebroski, H. W. Alter and F. K. Heumann J. Am. Chem. Soc. 73 5646 (1951)

Z4 N. J. Zinevich and L. A. Garmash Zhur. Neorg. Khim. **20** 2838 (1975)

Z5 V. L. Zolotavin Zhur. Obshchei Khim. **18** 813 (1948)

Z6 A. V. Zotov and Z. Yu. Zotova Geokhimiya **1979** 285

Z7 A. V. Zotov and Z. Yu. Zotova Geokhimiya **1980** 768

Z8 O. E. Zvyagintsev and S. B. Lyakhmanov Zhur. Neorg. Khim. **13** 1250 (1968)